CAMBRIDGE STUDIES IN ECOLOGY

Dynamic Biogeography

Dynamic biogeography

R. HENGEVELD

Institute for Forest and Nature Research
Arnhem, The Netherlands

CAMBRIDGE
UNIVERSITY PRESS

CAMBRIDGE UNIVERSITY PRESS
Cambridge, New York, Melbourne, Madrid, Cape Town,
Singapore, São Paulo, Delhi, Tokyo, Mexico City

Cambridge University Press
The Edinburgh Building, Cambridge CB2 8RU, UK

Published in the United States of America by Cambridge University Press, New York

www.cambridge.org
Information on this title: www.cambridge.org/9780521437561

First published 1990
Reprinted 1992
First paperback edition 1992

A catalogue record for this publication is available from the British Library

Library of Congress Cataloguing in Publication data
Hengeveld, R.
Dynamic biogeography / R. Hengeveld.
   p.   cm.
Bibliography: p.
Includes index.
ISBN 0 521 38058 8 (hbk) ISBN 0 521 43756 3 (pbk)
1. Biogeography.   1. Title
QH84.H46   1990
574.9–dc20   89–36092   CIP

ISBN 978-0-521-38058-4 Hardback
ISBN 978-0-521-43756-1 Paperback

To my parents

Contents

Acknowledgements

During my work on this book, several people have been important, either specifically by reading and commenting on the text, or by creating the necessary working conditions. Initial drafts, still written in Dutch, were read by Dr Max Van Balgooy, whose early interest was indispensible. Early English drafts were heavily criticized by Dr Jeremy Holloway and Dr Colin Prentice, who both told me much about what to include in a book and what not to include. Jeremy also suggested that I should go to the Anti-Locust Research Institute where at only the second visit the penny dropped, opening up the field of broad-scale movements of species and range dynamics. As recent insights in climatology are central in this field, earlier discussions with Colin about Bryson's work became relevant, as well as those of present-day palynology and quaternary research in general. Later, Professor Tom Webb commented on Part I and approached the text from yet another angle. Later stages, up to the last one, were read by Dr Francis Gilbert and Professor John Birks, who both clarified the text considerably. In particular, John Birks read many versions, making innumerable English corrections and putting nasty remarks and questions for me to solve in the margin. He was exceptionally patient in accepting all my alterations and additions that reflected the growth of my ideas. I am still surprised that he recognized its potential at such an early stage of development. I can only hope that I have done enough to make his effort and time investment worth the result.

The work was done while I was working at various laboratories in succession. Professor Victor Westhoff, in fact, suggested that I take on the work, though at that time it was still intended for a series of Dutch texts. Later, Professor Marinus Werger, Professor Jan Koeman, and Professor Rudolf Prins stimulated me with their interest. The librarians of several institutes had a heavy task with obtaining the literature, and

Ruud Wegman again did a good job with drawing the figures. Finally, much time, energy, and material costs, otherwise available for family life, have been given to this book, thus making Claire's task even heavier than it normally would have been.

Apart from these more tangible conditions, my parents should be mentioned in a more general way, as they conditioned me intellectually. I feel that it is their combination of theoretical interest, practical inclination and skill, and humanistic idealism that has shaped me. To a great extent, it is their influence that has thus given rise to this book.

I hope that this book as the combined result is such that its users keep the impact of the help and stimulus of all these people just as much in mind as I do myself.

Preface

When my friend, Victor Westhoff, asked me some years ago if I would write a book on biogeography, I hesitated. Two reasons encouraged me to try: (1) frustration that biogeographical classifications are rarely interpreted, and that any such interpretations are not tested, and (2) realization that two papers with Jaap Haeck on the distribution of abundance within species ranges (Hengeveld and Haeck, 1981, 1982) could give a new perspective to many biogeographical phenomena. Working my way through several problems, I felt I had generally succeeded, until another friend, Colin Prentice, suggested adding a final chapter on what future biogeographical work I thought needed to be done. I realized I had not achieved my desired unifying interpretation. I started all over again!

This book is the result. It contains a general introduction to the methodology of biogeographical classification, and presents ideas about the dynamic structure of species ranges. I emphasize climate as a basis for explaining many biogeographical patterns; the book has an ecological bias, which many biogeographers might consider a drawback. I do not, but it is up to the reader to judge if I have been fair to other approaches. The book is very much a personal synthesis.

I do not give pure theory, nor burden the reader with formulae and their derivations. If my approach proves to be fruitful, such theory could be added later by others when sufficient data are available to test more refined models than is currently possible. I feel that existing knowledge about biogeographical patterns and processes does not justify any mathematical sophistication. My approach is based on the wealth of empirical data, old and new, concerning distributions of organisms. Dealing with statistical procedures for processing large data sets, I try to emphasize the reasoning behind data analysis and to avoid writing a 'cookery book' for solving individual problems. Covering a large part of the most significant

biogeographical problems means that many text figures and references
had to be weeded out to keep the text within limits. I hope that my choice
has not severely limited the possibility of checking my inferences and of
further reading.

By concentrating on spatial biological processes, I try to unify some of
the diverse and apparently unrelated phenomena within biogeographical
research, and hopefully to clarify part of the present biogeographical
scene, and to assist in devising sharper and more precise pictures.

Introduction

Dynamic biogeography concerns the study of biological patterns and processes on broad, geographical scales and time scales. These processes can often be understood from processes operative at finer scales, or even from the properties of the individuals within species. In dynamic biogeography, therefore, it is as if one looks down onto the earth from a great height, and at first distinguishes broad-scale patterns and processes. Only later, on closer inspection and only insofar as they help to explain broad-scale phenomena, can one also identify fine-scale patterns and processes. Dynamic biogeography differs from studies of spatial patterns and processes that concentrate on fine-scale phenomena as seen in the perspective of broad-scale phenomena. This latter approach is distribution ecology, or, by analogy with climatology, synoptic ecology, where one looks in the opposite direction, i.e. from local through geographical to global scales. By presenting these two disciplines, dynamic biogeography and synoptic ecology, in this perspective, they clearly represent two approaches to the same phenomena, namely spatial patterns and processes discernable at different scales of variation.

However, these approaches are not separate or contrasting disciplines, and sharp dividing lines cannot be drawn. Instead, it should be realized that any distinction between biogeography and ecology is simply one of scale over which a phenomenon occurs, together with the direction in which one looks. At the operational scales of ecologists and biogeographers, the subjects merge into one another. The distinction between them is artificial and subjective, and potentially hinders our understanding of important spatial biological phenomena.

Although the immediate aim of dynamic biogeography is to describe and explain spatial patterns and processes of taxa, its ultimate goal is to aid our understanding of evolutionary processes. We must integrate information of a great number of properties of individuals, populations,

1

and the species as a whole. Thus, it will sometimes be necessary to collect information about an individual's physiology, anatomy, or morphology, whereas at other times information is required about the spatial distribution of genetic traits, behavioural characteristics, or population dynamic statistics. Biogeography should no longer be subsidiary to taxonomic classification, as it has been for such a long part of its history. Instead, it should develop its own problems, methodology, and models, and collect its own data. Yet it must derive many parameters of its models from those in other biological disciplines. In other words, we must formulate our problems so that biogeography acquires a firm place among other biological disciplines. Looking from the viewpoint of the central issue of biology, namely evolution, distribution ranges and the processes occurring within them should not be considered merely as a taxonomic character, but as the outcome of the interaction of a species' biological properties with each other and with the properties of the environment. Characteristics such as shape, size, structure, location and dynamics of species ranges integrate and reflect all these interactions, and can show their relative importance at a particular moment or through evolutionary time.

We still know little about these characteristics; although many hypotheses have been formulated about them, there are few data to test them directly. Biogeographers are often not particularly interested in doing this type of field work; they often leave it to ecologists. Ecologists, for their part, may feel unable to assist or are disinterested in unravelling a great variety of complex factors, operating on inconveniently broad spatial and temporal scales. They prefer to concentrate their analysis on restricted areas, laboratory experiments, or analytical or simulation models.

The lack of testable hypotheses cannot be explained by a lack of activity within biogeography. On the contrary, recent increase of interest in biogeographical problems indicates the opposite. What are currently needed are sound descriptive data, which at a later stage could be analysed to test hypotheses and theories. Sound and detailed descriptions of size, shape, structure, and location of species ranges are particularly lacking, and descriptions of dynamic aspects of ranges are practically non-existent. One can wonder why already existing data in regional and national atlases of European plants still await critical analysis, particularly since analysis of European pteridophytes by Birks (1976) has given such revealing results. The same can be said about maps of Scandinavian carabid beetles, European butterflies, and birds (cf. Hengeveld and

Hogeweg's (1979) analysis of Dutch carabid beetles in the Netherlands and Europe). It seems that despite the existence of plausible explanatory hypotheses, and despite occasional, detailed information on the regional occurrence of some taxa, biogeographical analysis is still not able to take off. What may be lacking is some formulation of more explicit and testable models on processes operative at the spatial scale of, for example, northwestern Europe. We also need to know more about what data-processing techniques are presently available, how they work, what they can solve, and how their results can be interpreted. This book attempts to supply this information, to develop a model of the structure and dynamics of species ranges, and to show the biological and ecological relevance of such studies.

Discussion of the dynamics of species ranges is an important and integral aspect, not only because it reflects a growing interest in processes as well as in patterns, but also because it is only by understanding processes that patterns can be explained. Species ranges are often not as uniform and static as the solid blots on maps suggest. Population density, for example, varies widely over the range, as well as from year to year. The same holds for other ecological parameters such as aggregation of individuals, habitat preference, and numbers of habitats they occupy. Changes in ecological conditions result in large shifts in local numbers of individuals and possibly in other ecological parameters as well, due to differences in net reproduction or to mass movements of propagules or adult individuals. Rainey (1978) uses the term 'oceanic approach' in the study of insects, comparing mass movements of individuals floating on air currents with those of marine species drifting in ocean currents, thereby covering great distances in a short time.

Looking at species ranges from this point of view not only introduces population dynamics into biogeographical studies, but also incorporates knowledge about species' habitat preferences, geographical and population genetical patterns, and their physiological and anatomical basis into this perspective, as well as anatomical and morphological properties connected with, for example, reproduction and dispersal. We can consider differences or similarities between closely or distantly related species in relation to such properties to explain differences or similarities in geographical distribution or range dynamics. Moreover, we can look at patterns in climatic fluctuation, either over short or long time periods to explain shifts in distribution or changes in range size. The statistical characteristics of ranges, such as size, location, internal structure, and dynamics, are different expressions of the biology of great numbers of living individuals considered together, and thus they should be studied as such. When conditions deteriorate, individuals or their descendants will emigrate to other, more favourable locations, resulting either in local shifts in density, or shifts in the entire range. When conditions change over great expanses of the earth, some ranges may expand, whereas others may contract or vanish. The result of continuous changes in environmental conditions at different spatial and temporal scales from fine, temporary, and local scales to broad, long-lasting, and global ones, is a permanent amoeboid creeping of species over the earth, and this in turn results in kaleidoscopic changes in patterns of species ranges relative to each other. The study of this endless movement, described in terms of ever-changing environmental conditions, is the heart of dynamic biogeography.

In attempting to describe the dynamism of species ranges, this book is arranged so that it gradually narrows our interest from the coarsest types of classification to the formulation of a dynamic model of species ranges. This is followed by the model's implications for understanding broad-scale patterns of coincidence shown by many species, as well as for individual species behaviour and future research in biogeography. This whole process, starting from the coarsest classifications and resulting in detailed models is divided into three major steps. The first step, described in Part I, comprises classifications in which no consideration is made of the identity of the taxa or their biological properties. Thus, from the information obtained, it is not possible to infer how the idiosyncrasies of a certain species enable it to fit into particular local habitats within its range, or how it maintains itself there in spite of, or due to, environmental dynamism. This is the field of biogeographical classification where biogeographical units such as kingdoms, regions, or districts are defined. It is the area from which concepts such as biogeographical elements are derived. This is also where much of the recently developed vicariance biogeography belongs, the proponents of which draw inferences from coincident distributions of selected taxa – those with disjunct ranges – in terms of phylogenetic development. These coincidences only emerge from classifications of these taxa. Although statistical testing is often precluded, we should use explicit methodology. This is why discussion of the methodology of biogeographical classification occupies four chapters in Part I.

Part II concentrates on classifications in which the identity of the taxon remains of little or no concern, but where the identity of certain biological properties is important. Here, the aim is to describe and explain continuous regional or global trends, or discontinuities in the frequency of one or more biological properties in space. Statistical analysis and its methodology is not emphasized here, although, as in the other two parts, large numbers of taxa that have (or lack) the properties concerned are analysed. Often, trends or differences in location of such patterns are so clear that statistical testing is unnecessary. As in Part I, identification of continuous trends or discontinuities in the distribution of some property serves to identify the ecological or historical factors operative at the spatial scale concerned.

Part III discusses classifications and biological phenomena for which knowledge of both the taxon and its biological properties is required. Statistical analysis of large numbers of unselected species is necessary to evaluate the generality of the patterns. Here, more than before, we are interested in patterns and processes within species ranges, rather than in supra-specific units or trends. Closer and closer looks at biogeographical phenomena result in considering processes on finer scales in time and space. This leads to an evaluation of both dynamic and stochastic aspects within the geographic processes of species maintenance through time. Together with spatially non-uniform distributions of the intensities of many population dynamic processes, these two aspects give insight into the nature of species ranges as dynamic response-surfaces relative to an environment that is heterogeneous and dynamic over all spatio-temporal scales.

By gradually concentrating on finer-scale patterns and processes, we may easily fall victim to reductionism by losing sight of the problems posed initially. Yet, as species respond to a variety of factors on all scales, information about finer-scale processes should not only help us to understand finer-scale patterns, but also to elucidate larger-scale ones. In Part IV, I pick up some of the threads of Part I on concordant patterns and explain them by processes described in Part III. Although this is only a first attempt at integrating multiscale phenomena, I try to indicate the need for their integration. Only by integration of various approaches, aspects, and phenomena can biogeography move from its anecdotal phase and acquire the status of an independent, mature biological discipline.

1

Topics

I define dynamic biogeography as the analysis and understanding of spatial biological phenomena in terms of past and present factors and processes. This definition implies five important topics. First, that we know what is meant by biological factors and processes compared with non-biological ones. Second, related to this, we must give biogeography an evolutionary outlook, that is we emphasize the continuous process of adaptation to ever-changing conditions on earth. Evolution is change, not only random change, but change for survival. This can be accomplished in two ways, by genetical adaptation and by spatial adaptation; the latter is of particular concern here. In this context we must be able to distinguish historical components within present-day patterns from ecological ones, consisting of factors and processes operative today. Third, at present, biogeographical methodology is still mainly inductive, which makes it important to adopt explicit, inductive reasoning. This implies that processes studied in dynamic biogeography are essentially statistical and are defined statistically. This means, among other things, that we must decide how far we are concerned only with patterns common to most species, and how much attention should be given to extreme or deviating phenomena. Such extremes can shed light on the interpretation of more general patterns. Another problem is whether we should model null-hypotheses in statistical terms only, or also in biological ones. Fourth, confining our interests to a certain aspect, approach, or process, restricts the possibility of generalizing our results. They may thus give a biased picture of biogeography. Confining our interest is often the only practical way to approach some problem, but we must be aware of the total framework of biogeography to which our results relate. Finally, processes are defined for the spatio-temporal scale at which they occur, as well as by the taxonomic level of the taxa concerned. Looking at phenomena occurring at one particular spatio-temporal scale or taxo-

7

nomic level implies using a filter, which prevents us from seeing phenomena at other scales or taxonomic levels.

We shall now discuss these topics in detail.

The biological approach to biogeography

Biogeography is a biological discipline, concerned with the biological phenomena. Yet, it is not always clear in what way we make it a biological discipline that differs, for example, from physical, geological, or mathematical ones. It is usually all too easy to present explanations that neither account for biological properties of the species concerned, nor explain them. For example, a small predator cannot eat a large prey, thus restricting the predator's diet. Such considerations do not lead to biological theories. But the predator's search for particular prey, possibly resulting in the same restricted diet, leads to the biological theory of dietary specialization of species. The problem is then whether the result is a biological theory, or a physical, geological, or mathematical one applied to biological data. If so, we should not speak of biological theories, but rather of physical, geological, or mathematical theories, possibly – but not necessarily – explaining a biological phenomenon. I suggest that biological theories should consider the origin and consequences of biological properties of species or of taxa in general. Two related aspects are relevant – (1) the specificity of biological adaptations and (2) the problem of whether or not general laws can be formulated.

In MacArthur and Wilson's (1967) equilibrium theory of island biogeography species number is related to island size, its distance to a nearby continent or to other islands, its age, and so on. This approach differs from that of Lack (1976) or Carlquist (1974), who both emphasized biological idiosyncrasies of island species and their adaptations to insular life. Similarly, Mueller-Dombois, Bridges and Carson (1981) considered the central issue of which environmental conditions on islands make them differ from continents. If they differ in certain respects, we may ask to what extent are species adapted to particular conditions, namely how do these species fit into them, or even how do they utilize them. Certainly, not all species have identical opportunities to colonize an island or to live on it for some time. Differences in these opportunities depend on the degree to which their specific properties match local conditions. Furthermore, the build-up of species numbers does not simply follow topological rules such as the degree of isolation or size of islands. As shown by Flenley and Richards (1982) for Krakatoa, this build-up follows successional steps of local vegetation development. Species may arrive too early, on time, or too late, depending on their biological requirements that do or do not match local conditions. If they arrive on time, they may stay for only a restricted time, depending on the vegetational development. The study of properties that influence why a particular species does or does not occur on an island can thus provide biological insights. However, its biology remains obscure when we just count species numbers, calculate species turnovers, and derive immigration and extinction rates. In those cases, species remain unknown, and their biology seemingly irrelevant. Viewed this way, MacArthur and Wilson's (1967) theory defines some boundary conditions within which biologically relevant processes occur. The study of these processes starts where the equilibrium theory ends.

Their theory was intended to be of general applicability to many different islands. But biological idiosyncrasies of insular species can rarely, if at all, gain a similar, and apparently desired, degree of generality (cf. Kareiva and Odell (1987) for a similar opinion after applying a mechanistic model in an ecological context). This poses another problem: are there general laws in biogeography? Without wanting to go too deeply into scientific philosophy, I suspect that many existing general biological laws, including MacArthur and Wilson's (1967), result from constraints imposed by non-biological laws on biological variation, and that, as a consequence, they are of biological interest only as constraints.

Allometric-growth relationships may be viewed as one of a few, or even the only, general biological law (Peters, 1983), putting constraints on a species' physiology, anatomy, behaviour, and ecology. In my view, mechanical considerations within this context are biologically interesting only in so far as they make understandable different adaptations to the problems imposed by these constraints. Thus, Reynolds (1984) showed that the mucilaginous algae do not conform to the mathematically expected relationship between body size and volume, making them better adapted at remaining in suspension than non-mucilaginous species. To the same end, they can also form colonies, or have irregular shapes. These adaptations, their origin, and their effects make allometric-growth relations biologically interesting, rather than the law itself which results from the general applicability of chemical and physical constraints.

Thus, distinctions made between oceanic islands and continental ones because of their geological history have nothing to do with biology, as it is impossible to predict which species are found on either of them, or what adaptations to local conditions these species might have. The same holds for theories based on continental drift, geomorphological properties of islands, general models of species diffusion, or climatic and vegetational classifications in zonal or azonal distribution patterns. In such cases, species are usually assumed to be the same. This is not true, and is contrary to the essence of biology as explaining different forms of life as specific adaptations to environmental conditions. Moreover, suits of traits, together forming multivariate ecological response syndromes, prevent the formulation of general laws. Just as in population genetic breeding systems, we can well distinguish various combinations of traits leading to the same end and forming physiological and demographical systems.

In later chapters I try to explain biogeographical patterns and processes in terms of species properties. To this end I will discuss those techniques that extract information from biogeographical observations for subsequent model building.

The evolutionary approach to biogeography

Viewing biogeography as a biological discipline means not only that we describe geographical patterns and processes, but also that we place them in an evolutionary context. This is not to say that we must try and straitjacket all phenomena or even our approach in this way, as it will often not be necessary to consider evolution or, more particularly, speciation. Essentially, evolutionary change is only a spin-off of adaptation to changeable conditions. The concept of evolution, as used here, is a tool of biogeographical research; it is a means to an end, not an end in itself. We assume that evolution has occurred in order to understand biogeographical patterns and processes. Within a biogeographical context these patterns and processes are not used to explain evolution, as this would shift emphasis away from biogeography to evolutionary biology. Both can be studied in their own right, as shown in later chapters. The time-scale of field investigations is such that evolutionary adaptations can hardly, if ever be observed. If the time-scale chosen is long enough, sampling intensity will usually be too low to give a reliable picture of the patterns and processes operative in the past.

By adopting an evolutionary approach one assumes that every species (1) adapts to conditions in its environment in its own, individualistic way, (2) at present it is adapted just as it has been, since its origin, and (3) its response to environmental conditions becomes less homogeneous throughout its range as the physiological responses of all species resemble each other more. We will briefly look at these assumptions in turn.

The first assumption is that species not only adapt in space to changing conditions by shifting, contracting, or expanding their range, or by redistributing individuals within it – that is by purely biogeographical processes – but also by shifts in population genetic content. The result of these two processes together, spatial adaptation and population genetical adaptation, is that species closely relate to their environments. This fit is never perfect, but it will leave scope for further environmental change without necessitating repeated species adaptation. Species with different physiologies and morphologies are usually found side-by-side. Yet, as a consequence of evolutionary changes, all species have their own habitat requirements, due to differences in physiology, morphology, and other properties. These differences make them unique entities. Moreover, species do not respond to changes in their environment as parts of a particular community, but, due to their unique set of biological properties, they behave individualistically. Thus, sometimes species may live side-by-side, at other times they separate, going their own way at their own speed and in different directions (e.g. Davis, 1981a; Graham, 1986). We can expect the same individualistic behaviour relative to congeneric species. The occurrence of many such species in a small area, so-called species nests, does not necessarily reflect their area of origin, but simply the degree of similarity in their ecological responses.

The second assumption is that species are adapted to the same environmental conditions that existed when they originated. They have always had the same general habitat requirements, and their life histories and dispersal capacities have always matched, more or less, similar patterns of environmental fluctuation. It thus makes little sense to study geographical patterns and processes as if they were generated or maintained in isolation from the species' environment, rather than reflecting it. Yet the bulk of historical biogeography does not consider the ecological requirements of particular species, the environmental conditions that the species encounter, or possible changes in these conditions. It is only recently that the study of the causation of extinction processes has begun to be considered and the study of environmental factors operative during speciation is in its early development, as is the study of factors that induce species migration. Adopting an evolutionary viewpoint through ecological adaptation is thus a first step in understanding geographical adaptation of species in relation to a continuously changing environment.

Finally, assuming that species adaptation occurs relative to environmental conditions implies that the numbers of individuals are not uniformly distributed within a species range, but that they show statistically an optimum distribution with highest numbers in the range centre and lowest near the range margin. Likewise, other population-dynamic parameters are distributed in the same way, reflecting a species' response to geographical trends in intensity of environmental conditions. These spatial distributions, measured over a certain period of time, can either be considered chance distributions or distributions of the intensity of occurrence. As chance distributions they represent the geographical distribution of the local probabilities of developing a certain density, which, in their turn, depend on many population-dynamic parameters and processes including dispersal, as well as on the frequency distribution of favourable and unfavourable conditions. They represent the dynamic component of these distributions. As intensity distributions they represent the intensity of physiological processes such as growth and fecundity, conditions which are most optimal in the range centre, becoming more and more unfavourable towards the margins. They represent the static component of these distributions.

Of course, a sharp distinction between dynamic and static components cannot be made, as they merge and depend heavily on one another. But the distinction shows clearly that we must integrate information from several biological disciplines to understand species ranges, dynamics, and development through time. To understand species dynamics we require population-dynamic parameters as well as population genetic ones, both of which are rooted in a species' genetical, physiological, anatomical, morphological, and behavioural properties. To understand their statics we must understand their physiological properties at the outset. Adaptations involving these properties determine whether or not a species can survive at some place or migrate fast enough when conditions deteriorate. All adaptations evolve, and differences from species to species determine differences in species' occurrences and behaviour. Without considering a species' biological properties, we can hardly understand its biogeography, if at all. An evolutionary outlook allows us to view the adaptive value of species properties, as reflected by range structure and dynamics. This also applies to biogeographical differences between species. It integrates knowledge from many different biological disciplines at the geographical level of investigation.

The inductive approach to biogeography

As biogeographers often infer their explanations from existing data, biogeographical methodology is mainly inductive. Island biogeography is an exception as it gives a deductive, analytical model from which future observations can be designed or experiments based.

However critical this distinction may be in theory, in practice it is not simple to adhere to it for two reasons. Firstly, parameter choice in any analytical model depends on inductive reasoning. Straightforward parameter-derivation from existing models in other parts of biology is far away, and it is doubtful if it is at all possible in biogeography. Secondly, hypotheses or models of some phenomenon must exist before we can test its existence or operation. The first question about any phenomenon, prior to any suggestions concerning explanations, is always whether the phenomenon exists. Testing this requires knowledge of alternative processes which is rarely available.

Diamond's (1975) assembly rules are a case in point. Both biological and stochastic processes result in spatial variation, which can be characterized by a probability distribution. In Diamond's case these distributions are represented by the marginal totals of the matrix of species presence–absence of birds on islands and they differ for different biological and stochastic processes. They can form a uniform distribution if, because of interspecific competition, the species are evenly distributed (Diamond's (1975) checkerboard distributions). When they are randomly distributed, the marginal totals could be binomial, lognormal, Poisson distributed, or more complex, depending on the processes concerned. Testing if bird distribution among islands differs from random expectation thus requires analytical construction of a probability distribution from which random samples are assigned into the cells of the matrix of expected observations, after which actual observations are compared with expected ones. Analytically deduced probability distributions have advantages of greater theoretical flexibility and avoid circularity, which marginal totals do not (Diamond and Gilpin, 1982). But their application is difficult because of present ignorance of the processes involved, leading to the one or the other probability distribution, which may explain why Connor and Simberloff (1979) used existing data from which they derived expected cell contents (see also Snijders, 1984).

An excellent example of testing geographical processes concerns the distribution of winged and unwinged carabids over islands east of Sweden. Ås (1984) assumed that winged species are airborne, and unwinged ones waterborne, implying that their probability distributions of arrival follow exponential and normal distributions, respectively. The exponential distribution only relates to distance to the mainland, (winged) individuals declining with a constant rate, independent of their biological behaviour. The normal distribution, in contrast, relates to the physical properties of rafts that (unwinged) individuals are thought to be dispersed on across the sea. When rafts continuously change their course, and their direction follows a normal distribution, the individual's probability of arriving on a certain island is similarly distributed. Next, Ås calculated proportions of expected arrivals of winged species versus numbers of unwinged species at various distances from the mainland, assuming an equal emigration rate. No biological property needs to be involved to explain this distribution additional to being airborne or waterborne as the proportion of winged versus unwinged species over the islands is not statistically different. This study shows how probability distributions chosen inductively were used to deduce the spatial distribution of species from a few dispersal characteristics.

The difficulty is that not only do theories depend on observations, but that observations also depend on some theory. Science is a bootstrapping process. In this process, classification or ordering of the data according to a subjectively chosen criterion is the first step. The next is to interpret the patterns obtained, followed by testing of inferences drawn from this interpretation. Gradually, after many attempts, a pattern may emerge that is common to all and that, inductively, leads to testable models and hypotheses integrating all available information. This allows, as a next step, more systematic data to be collected for testing the models and hypotheses constructed analytically. Only then can we apply a deductive methodology directed to obtain insight into the model's properties relevant to the biology of the species concerned.

I feel that biogeography greatly needs means to bring scattered and often anecdotal information together. Much attention should therefore still be given to classification or ordination, their methodology, and to the results obtained. Together with information from, for example, physiology, morphology, ethology, genetics, ecology, and climatology, these results will eventually allow us to formulate explanatory models for the structure and dynamics of species ranges, thus demanding sound deductive reasoning.

Biased approaches to biogeography

Biogeography, like other sciences, is heterogeneous in its statistical-testing procedures and deductive methodology. Although we cannot presently perform statistical tests in many cases, we must base our conclusions on large data-sets. For each species many individuals are needed and for patterns among species many species are also required.

Hypotheses about geographical patterns and processes can never be generated from a single individual or species. In this regard, biogeography is similar to taxonomy, recalling Mayr's (e.g. 1963, 1982) insistence on population thinking, as opposed to typological thinking, in essence describing collections of individuals in terms of parameters as the mean as well as variance, kurtosis, and skewness. Just as in biogeography, these parameters were not used for statistical testing of hypotheses. Data selection always imposes constraints on generalizing results beyond the collections from which the results originate.

Vicariance biogeography is one such case, inferring generalized tracks from several common connections drawn between parts of disjunct ranges of supraspecific taxa; the historical interpretation of these tracks cannot always be generalized for species with continuous ranges. (In this context, tracks are not meant to be dispersal routes, but lines on a map, drawn between remaining parts of a fragmented area.) Restricting interest only to those cases where diffusionary dispersal falls short (as in vicariance biogeography), or where it is of overriding importance (as in island biogeography) narrows our view about biogeographical processes (Pielou, 1981). Inadequate or predominant dispersal are extremes in a range of variation in dispersal, and, as such, occur mainly in extreme situations. Hence, vicariance biogeography often concentrates on topological effects on distribution patterns from shifts in continental positions, whereas island biogeography is mostly applied to offshore islands. Application of vicariance biogeography to island archipelagos can raise difficulties (Holloway, 1982), just as island biogeography is of limited value when applied to large, ecologically diverse continents, unstable islands (e.g. Lynch and Johnson, 1974), or species differing in their immigration and extinction probabilities (Strong, 1979). To understand biogeographical patterns and processes not only among islands or continents, but also within them requires a more comprehensive approach (cf. also Thornton, 1983).

Restriction to historical or ecological processes also leads to an incomplete understanding of biogeographical processes. Ecological processes quickly reach a new equilibrium condition, whereas historical processes concern a certain time period or time-lag before this equilibrium is reached. Thus, historical processes are defined here in terms of both ecological factors and factors causing a time-lag, such as low natality, mortality, or dispersal capacity. Current biogeographical processes that also operated in the past are termed palaeobiogeographical processes, together with their resulting patterns that are studied in palaeobiogeography. As vicariance biogeography exclusively relies on historical patterns and processes, and island biogeography on chance processes, they describe only part of the geographical processes. They leave untouched that part of the patterns that result from ecological processes, although it is these that ultimately define a species' survival probability (cf. also Endler, 1982).

As collections of selected patterns, processes, and scales of variation are restricted because of the selection procedure applied, their explanations thus have restricted applicability as well. To make generalization possible, I mainly use studies on large data collections that permit statistical descriptions of species and of individuals within species. This, in turn, allows inferences to be drawn about possible explanations of distribution patterns that can be tested by further research.

Scales of variation

Above, I have mentioned problems that may arise from differences in the scale of variation, both in time and in space. For example, two patterns, a fine-scale one and a broad-scale one, are superimposed, similar to daily temperature fluctuations superimposed on seasonal ones. However, the ratios of two fluctuation patterns often differ from such a picture, nor are they fixed as in daily and seasonal temperature variation. Both the amplitude and period of fluctuation of the fine-scale pattern may be smaller, for example, whereas the broad-scale pattern remains the same. The first may even be so small that it is considered irrelevant noise. Conversely, we can increase the amplitude and period of fluctuation of the fine-scale pattern relative to the broad-scale pattern, eventually making them equal to those of the broad-scale pattern. Thus, scales of variation are neither absolute nor discrete phenomena, but they merge into each other, depending on their relative characteristics. They form a continuum. Only in certain parts of temporal variation are they discrete. That scales of variation form a continuum does not, however, mean that plant and animal species are uniformly distributed; in fact, they are usually clustered. This clustering can be enhanced by or can even originate from interactions with other species (e.g. Heads and Lawton, 1983; Kareiva, 1986). But we cannot separate out certain or even fixed scales of variation *a priori*. In that sense only, they form a continuum rather than being discrete.

They are only rarely discrete, and are usually variable. When, for example, a plant species is dispersed over a flat meadow in which only one factor, say soil moisture content, varies, the plant will either occur uniformly over this meadow, or not at all. If the meadow slightly slopes down, the species will occur only along a certain part, its densities following a unimodal, optimum-response surface. Variation in slope angle thus affects its local area or range (or period in temporal terms) and hence the species' scale of variation. Because the steepness of slopes varies continuously, variation in species dispersions is continuous as well. Apart from variation among fixed slopes, local disperson also depends on variable factors such as precipitation, soil condition and penetrability, causing species to migrate to other locations in dry or wet years. However, despite similarities, geographical ranges cannot always 'move with the temperature', but depending on responses to other factors, they can decrease or increase when temperature varies. Thus, tree species which occurred widely during the Tertiary, are found today as local endemics due to lower temperatures.

Spatial variation also differs in other respects than continuity and changeability. Temporal processes vary in one direction only; seeds germinate, and young plants grow, flower, produce seeds and die. Within certain limits the next step can be predicted from the former. This differs in spatial variation, because the relevant factors usually vary at all rates and scales and in various directions. Knowing the environmental conditions or the biotic composition at one locality hardly tells anything about other sites. Only on broad, global scales can patterns be recognized parallel to latitude, or longitude, but these patterns are not the only ones of biogeographical interest. Why a species occurs in a particular part of a continent or ocean, and why its range has certain dimensions and spatial dynamics usually depends on local conditions, which are less predictable from one locality to another. Range location, structure, size, and dynamics to a large extent depend on factors varying on different scales and in different directions.

It is sometimes assumed that temporal and spatial variation are positively related (Birks, 1986; Delcourt, Delcourt and Webb, 1983). However, this is not a general rule. For example, man's impact on nature, though short in geological terms, has been disastrous for many terrestrial and marine biotas on a world scale. The same holds for invasions (e.g. Elton, 1958) and epidemic diseases, such as Dutch elm disease in Europe, or chestnut blight in North America. In a few years, such diseases may spread over one or more continents. Too little is known about the geographical history of the majority of species to generalize about this supposed relationship.

Two more factors make the picture of daily fluctuation superimposed on seasonal fluctuation too simple for generalization. Firstly, it suggests a nested or hierarchical structure for environmental factors and processes and for the resulting biological phenomena. However, nested or hierarchical structures are not always clear, nor are they constant. Moreover, the importance of environmental factors often depends on other factors. Also, changes in their relative importance affect a supposed hierarchical structure of scales of variation. Species ranges in the northern hemisphere may thus be limited in the north by too high moisture conditions and in the south by too high temperatures (e.g. Gauslaa, 1984).

Secondly, the above picture of variation on different scales suggests that fine-scale variation is likely to be less important relative to broad-scale variation. However, demographic processes, for example, based on various biotic interactions among individuals – and hence restricted to short distances – may keep numbers over the whole geographical range low. Fine-scale processes may then result in broad-scale effects, which, consequently, do not fit the hierarchic conception of scales of variation. Also, for long time-scales the pattern may be erratic, although it was regular on a shorter scale. Thus, on fine scales genetic drift may dominate population genetic variation, whereas selection factors may show up on longer temporal scales. Yet, on still broader, geological scales the pattern may be erratic again and genetic drift is the most appropriate model (cf. Lande, 1976), when selection changes direction from one period to the next.

The consequences of recognizing differences in scale of variation are manifold. One consequence pertains to sampling and another is that generalization from small, local sampling areas to vast species ranges, or even latitudinal trends is not possible. Also, knowledge of the main factors determining a species' altitudinal distribution cannot be extrapolated to range delimitation or structuring, causative environmental variables differing in various parts of the range. Moreover, other factors, such as differences in migrating distance, or in day length, can also show up in geographical space, whereas they do not altitudinally.

It is revealing to compare results from analyses on two or more scales, because patterns and processes can differ on different scales. For example, Kikkawa and Pearse (1969) compared Australian distribution patterns of bird species with generic patterns. As the species were supposed to reflect present-day conditions, whereas the genera would reflect glacial conditions, these authors reconstructed historical patterns and processes leading to present-day conditions. Moreover, one can also compare the patterns found with the species' physiological preference, their ecological preference, and their geographical distribution (cf. Hengeveld, 1985*b*, for the development of the Dutch ground beetle fauna during the twentieth century). Such spatial and temporal comparisons result in definitions of geographical and historical elements *sensu* Wulff (1943).

Finding no effect of some factor, either locally (say temperature) or geographically (say competition) does not mean that it has no effect, but it may indicate that on the scale investigated it cannot be discerned and its effect is constant or uniform; observations at another scale of variation might reveal different results. Not only is extrapolation from patterns at one scale to those at others uncertain, but so is that of results from laboratory experiments to the field; finding a relationship between temperature or humidity and a species' demographical parameter does not necessarily explain its local abundance in some part of its range. This problem is more severe when the effect of the operation of some variable is on another scale than the operation itself. Usually, the spatio-temporal scale of effects expands with distance in time from the cause of a process. This not only gives problems in sampling, analysis, and interpretation, but also burdens processes of species interactions.

One example, summarizing scale-dependent variation, concerns the observation at some laboratories that after two decades of ecological and selection experiments *Drosophila* flies did not escape from their bottles, whereas they did before. This shows that (1) over the years and independent of several heavy short-term mortality and selection factors the occasional escape of flying individuals represents a selection – and, ecologically, a mortality – factor, having operated on a much longer term; (2) concerning the many successive generations during that time, the intensity of this factor was, as a selection factor, almost immeasurably small and as a mortality factor insignificant, which (3) could be detected only after, not during, the process happening. One or more of these three elements are, as we shall see, present in all ecological, biogeographical, historical, and geological processes. Moreover, they make it theoretically impossible to distinguish these as separate disciplines, however useful their distinction is in practice.

Apart from technical and interpretative difficulties related to differences in the scale of variation, these differences can also cause difficulties, or occasionally opportunities, to the organisms or species themselves (Hengeveld, 1987*a*). For example, it has been suggested that larger individuals by being numerically more stable have a greater chance of survival than smaller individuals, as they are less at the mercy of small-scale random fluctuations. This would also hold for, for example, home ranges, territories, or for annual, biennial, or perennial plants, the seeds of which may also germinate the same, next, or after several years, often all within one species. As environmental variation is not only capricious, but also operates on different scales, species should be adapted to variation at several scales simultaneously.

It is thus important not to live momentarily, but, so to speak, to look ahead. Time-lags are usually negative relative to time, but they can also be positive, the former being known as historical factors and the latter as anticipation. Biological systems often deal with both, adaptations often stabilizing the species' chance of survival by extending the time-scale in both directions. Thus, dispersal mechanisms may be seen as structures of spatial anticipatory adaptation, similar to mental abilities of forecasting events. It is understandable that the first plants had large, dyschorous seeds and that later plants and insects evolved more dispersive seeds or life stages (e.g. Schuster, 1976; Kennedy, 1928, respectively).

Results must relate to the scale of observation and explain the spatio-temporal scale of variation of the pattern or process observed. Distinctions made between historical biogeography, palaeoecology, physiological biogeography, and ecological biogeography are senseless when their results are not fitted into one perspective. To understand a species' geographical behaviour, it is usually necessary to know its temperature requirements for growing, for seed germination, or whether or not it has stolons. Reproductive capacity may be relevant in some cases, as is the ability to fly or to be blown away. In some instances geographical factors usefully explain the dynamic behaviour, locality and size of range; in others history or morphological properties are most important. But always all these factors or processes must be integrated in any understanding of survival. Biogeography studies where and how a species' individuals match geographical properties such as soil factors, physical barriers, or the dynamic intensity distribution of climatic elements. Any restriction to one or other scale of variation restricts our view, if not it distorts our understanding.

Conclusions

Species distributions must be approached from the viewpoint of their idiosyncratic properties, which all add to their survival in particular parts of the earth and not in others. It is the biological process of spatial adaptation of individual species that should be described. Certain properties dictate a smooth spatial adaptation, whereas others put species into difficulties or even lead to extinction. The integration of biological idiosyncrasies, reflected in a species' behaviour in geographical space will concerns us in this book; the techniques and concepts used are tools towards this end. As the phenomena concerned are of populations, species, or groups of species, they are described by numerical parameters, which necessitates using statistical methodology, techniques, and concepts.

Of course, studying taxa on all possible scales is impossible; one has to make a choice. This implies that a methodology should be used that minimizes the risk of selecting taxa, areas, or scales of variation and hence bias. One way is to start from the broadest scales, working down to finer ones, rather than the reverse, thus keeping the total number of degrees of freedom for the choices at a minimum. Another way is to apply hypothesis-generating techniques before starting more sophisticated types of analysis. Both approaches coincide in Part I, where I discuss classification of biotas at various spatial scales.

I

Patterns of concordance

Classification can be considered as the first but coarsest step in bio-geographical analysis. It should be done when little is known about distributional patterns and processes within and among species in geographical space. This view is basic to the present approach reflected in the structure of this book, which is just the opposite to that taken in many other texts. Indeed, biogeographical texts usually start with a discussion of the spatial distribution of species and the relevance of ecologically relevant parameters with respect to this distribution, and end with partitioning the earth into larger or smaller biogeographical categories. But Darlington (1957) correctly argued that such categories are not more than yardsticks to express the degree of overlap between the ranges of species or higher-order taxa. Such an overlap, or a certain deviation from it, can have biological significance and may give a convenient starting point to generate and test hypotheses. Viewing biogeographical classi-fication this way, it seems logical not to round a book off with its discussion, but rather to start with it.

First I discuss qualitative and quantitative approaches to biogeo-graphical classification, and then spatial categories or units as have been described in the last 150 years. After this, a recent example of the quantitative approach will be discussed. In Chapter 4 I consider recent criticisms of biogeographical classification as a preparation to the follow-ing chapters on the interpretation of biogeographical patterns of con-cordance. Instead of being an end in itself, classification can be used as a tool for generating interpretative hypotheses. Yet, in this, classification is only one means, and not necessarily the most efficient one. Another set of techniques, called ordination techniques, can also be used, and these can be more efficient. Therefore, Chapter 5 discusses a number of these techniques and compares ordination and classification results. This leads logically to the second part of the book, which presents a more detailed look at biogeographical phenomena involving description and interpretation of geographical trends in species richness and biological traits.

2

Qualitative and quantitative approaches

The qualitative approach to biogeographical classification partitions biotas intuitively into a system of groups, usually hierarchically nested. Here, the word 'intuitive' is not meant to be denigrating, as results of this approach often depend on sound knowledge of the distribution patterns of many species. Instead, I mean that the classification criteria are not made explicit. For example, one species is often considered more characteristic of a certain spatial unit than other species, implying potentially subjective and intuitive selection and weighting procedures. The difficulty of such procedures is not that their results are unreliable, but that other students do not know precisely how they have been arrived at, resulting in uncertainty. To avoid this, procedures were developed that specify each step. Thus, intuitive steps are replaced, as far as possible, by explicit procedures, specifying the nature of the various choices.

Woodward (1856) made one of the first attempts to delimit biogeographical units explicitly, proposing that for two areas to be considered different, at least 50% of their species should be different. The trend towards greater explicitness commenced soon after the turn of the century with Jaccard's (1908) similarity coefficient and got into full swing after the 1950s, particularly with the introduction of clustering techniques. This chapter discusses, with examples, the qualitative and the quantitative approaches.

The qualitative approach
The qualitative approach stems from Sclater's (1858) classification of the global bird fauna into several large spatial units. In later years refinements were made, although the spatial delimitation of the units and their hierarchical level were disputed. Also, opinions differed about both the number and naming of these levels. Van Balgooy (1971), for exam-

25

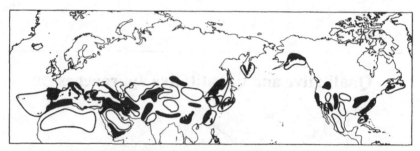

Figure 1. Areas of investigation for the classification of Lepidoptera along the southern border of the Holarctic (after De Lattin, 1957).

ple, published maps showing 12 classifications of the terrestrial Pacific biota, along with a variety of terms used for various hierarchical levels. Darlington (1957) gave a list of taxa characterizing the units he distinguished, but such lists, as well as descriptions of the ranges of these taxa, often remain unpublished. The resulting lack of basic information prevents subsequent checking of results, giving instability to biogeographical classifications.

Qualitative biogeographers did not stop at distinguishing several geographical units and arranging them in a hierarchical system; they also attempted to explain them causally, both in historical and in ecological terms. However, in the qualitative, intuitive approach one runs the risk of basing the biotic grouping unwittingly on causal factors. However, using causative factors as classification criteria prevents one from explaining the units so distinguished in either the same or in other terms. Although this may seem obvious, this methodological mistake is common. Particularly, though not exclusively, in the quantitative approach attempts are made to avoid this kind of circular reasoning, leaving the way open for subsequent explanatory research.

An example of the qualitative approach

De Lattin's (1957) analysis of Holarctic Lepidoptera (Figure 1) illustrates the qualitative approach of biogeographical classification.

The first step was to choose species with 'small or average' range sizes along the southern border of the Holarctic region, after which several 'primary' groups of distribution ranges were distinguished. Although each of the species chosen was restricted to one of these groups, a high degree of overlap was felt sufficient for group membership. Species overlapping with a single primary group were called monocentric species;

monocentric species partly coinciding with another primary group were called faunal elements. So-called polycentric species with relatively large ranges and therefore coinciding with more than one primary group were initially excluded from the analysis.

The second step was to partition each primary group into 'secondary groups', and these again into 'tertiary groups', and so on. Conversely, the primary groups were put together into three large 'macro-ecological' units, the arboreal, eremial, and boreal faunas. Although De Lattin recognized connections among these latter three units, he did not consider them real zoogeographic entities; instead they were felt to be of practical value only.

The third and final step was to interpret the groups historically. Since the location of the primary groups coincides with supposed glacial refugia, they were called 'expansion clusters'. Polycentric species, according to this interpretation, occurred in more than one refugium and their populations are not separated by spatial barriers.

The first two steps in this procedure are intuitive, as no criteria are given as to why and how to choose the Holarctic region, how to delimit it, how to delimit species with small and average range sizes from those with larger ranges, how large a coincidence between ranges enables recognition of primary, secondary, tertiary, etc. groups, whether or not groups overlap because the constituent species do not overlap completely within each individual group, and so on. Nor are the levels of primary, secondary, or tertiary groups defined exactly, although the system attempts to be general and independent of the taxon concerned. Moreover, the three macro-units are ecologically defined, whereas their interpretation is historical. No way of testing this interpretation or the biological delimitation of the groups is suggested.

Therefore, although De Lattin (1957) intended to criticize qualitative biogeographical classification, proposing his alternative step by step, he followed a procedure that, instead of being a criticism, illustrates the methodology and limitations of qualitative classification.

The quantitative approach

Woodward's (1856) criterion for judging the resemblance of two biotas was succeeded by Jaccard's (1908) more successful coefficient $s = c/(a+b+c)$ (where s is the biotic similarity, c the number of shared taxa, and a and b the numbers of taxa particular to either of the biotas) and which is still frequently used. The introduction of this and other similarity coefficients was a great step forward, but the problem remained how to assess patterns within many similarity values simultaneously. The number of values increases quickly according to the geometrical series $0.5n(n-1)$, where n stands for the number of areas compared. In Van Balgooy's (1971) classification of Pacific phanerogam genera, for example, n is relatively small (36), but the resulting 630 similarity values are too many to comprehend by eye. Cluster techniques reduce this information by giving a simple visual representation, a dendrogram or tree-diagram in which similar areas are connected at low levels or nodes, and more distantly related ones at higher nodal levels.

Statistical criteria for evaluating cluster techniques will not be given (cf. Pielou, 1984); two properties affecting the interpretation of clustering results must be mentioned briefly here. First, in principle, taxa from the biotas compared are weighted equally, although some procedures have the option to weight differentially according to some taxa or areas with particular relevance to the classification. However, differential weighting introduces information that ought to be used for explaining, rather than for constructing, classifications. It prevents the resulting classification being explained in terms other than those of the weighting criterion used.

Second, biogeographical units are either defined monothetically or polythetically (cf. Sneath and Sokal, 1973). Monothetic classifications define units by one or a few properties unique to them; they are defined qualitatively. In polythetic classifications the unit members share most of their properties, and the lack of one or a few is largely irrelevant; they are defined quantitatively on overall similarity. Their use in taxonomy, for example, implies that losing some property during evolution does not necessitate the classifier putting the species concerned in a separate genus or family, but generally it can remain in its original genus. In biogeography it means that taxonomic changes or species migration need not affect classifications dramatically.

Thus, by applying cluster analysis based on similarity values, the degree of biotic resemblance is visualized and made explicit.

An example of the quantitative approach

Analysing the distribution patterns of 242 North American mammal species, Hagmeier and Stults (1964) and Hagmeier (1966) considered biogeographic units as those having a relatively homogeneous fauna separated from adjacent areas by zones of heterogeneity. Generalizations are possible over relatively large areas to allow ecological or historical interpretations and predictions of range limits to be made. Because their definition of biogeographic units depends on spatial variation in faunal homogeneity as a criterion, they started with analysing this variation. To do this, Hagmeier and Stults (1964) defined an Index of Faunistic Change, $IFC = 100L/n$, where L is the number of range limits in 2490 squares of 50×50 miles square and n is the number of species in those squares. To estimate whether this index shows a clumped or a random distribution over North America, they took 10 samples of 100 IFCs each and estimated if the observed distributions differ from a Poisson distribution. The significance of this difference suggests that the IFCs are clumped and, hence, that classification is appropriate. They then partitioned the IFCs into 7 size classes and mapped their distribution over North America. Adjacent quadrats of a similar IFC class were then connected by isarithms indicating transition zones between areas of greater homogeneity. In the first study these limits were drawn so that they coincided optimally with existing classifications based on non-explicit methods, resulting in 24 'primary areas'; in the second they were not so adjusted, giving 86 such areas.

After defining primary areas, Hagmeier and Stults estimated the compositional affinity between areas by applying weighted paired-group cluster analysis. First, they calculated similarities using Jaccard's coefficient, defining a species' presence in a primary area when 50% or more of its range coincides with it. These similarities were put together in a half-matrix trellis-diagram, containing 276 values for 24 primary areas and 3655 values for 86 areas. Within this, the areas are arranged so that the highest similarities lie along the diagonal and the lower ones towards the corner. This arrangement follows from inspecting a dendrogram, summarizing the various biotic interrelationships hierarchically. Their final step in classifying the faunas was to define various categories according to certain similarity levels. Considering several criteria, they estimated a lower similarity limit of 62·5% for a province, 39% for a superprovince, 22·5% for a subregion, and 0–8% for a region, giving 22 provinces, 9 superprovinces, 4 subregions, and 1 region (Figure 2). Later, Hagmeier (1966) started from 86 primary areas and adjusted the similarity limits of the categories, giving 35 provinces with a similarity of at least 65%, 13 superprovinces (42·5%), 4 subregions (22–27%), and 1 region (0–8%).

To interpret this classification, the total number of species per province or the number of species characterized by a certain biological property was counted. Maps were prepared for numbers of species present per province, numbers of species per genus, numbers of

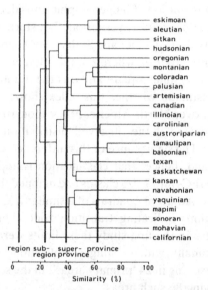

region sub- super- province
       region province

0   20   40   60   80   100
              Similarity (%)

Figure 2. Definition of various hierarchical levels in the classification of
North American mammals (after Hagmeier and Stults, 1964).

endemics, mean body size, dormancy, various life-forms, and geographical connections with other continents, thereby adding ecological interpretations of the provinces (Hagmeier and Stults, 1964).

Here we are not concerned with whether the classification is definite
and general or whether the interpretation of the provinces is correct.
Rather we concentrate on the attempt to make the assumptions explicit,
and to estimate the effect of changing certain assumptions. First, primary
areas are defined differently, for example (1) according to the number of
ranges, (2) adjusted to existing classifications or not, (3) according to
some subjective subdivision as for the Mohavian province, or (4) according to some physical barrier as with islands. As shown, adjusting to
existing classifications has profound effects on the number of primary
areas and hence on the classification of these areas. Subdividing a primary
area artificially or according to some physical barrier shows that such
faunas also differ at a provincial level. This makes one wonder which
criterion should be applied; a subjective or a physical boundary, or
primary areas delimited by the relative number of range limits, the latter
being dependent on quadrat size. Moreover, the criteria for delimiting
primary and secondary areas, such as provinces, are dependent, being

derived from the same model. The definition of species being present in a primary area when at least 50% of its range coincides with this area can also be modified. Finally, the choice of four spatial categories can differ, as can their delimiting similarity levels.

All these choices, although subjective to some extent, are made explicit. As we cannot avoid their subjectivity, nor the effects of other choices on the final results of the classification and its interpretation, we must make them explicit. Doing this allows different people to understand the results, which they otherwise cannot. If in disagreement, they can reanalyse the same data and can make their own choices.

Conclusions

Two broad types of biogeographical classification exist, the qualitative and the quantitative approach, both serving the same purpose. The quantitative approach makes explicit decisions which remain intuitive in the qualitative approach. The development of numerical techniques facilitates a replacement of monothetic classifications by polythetic ones, which affects classification interpretation. Weighting of taxa or biotas, though feasible in numerical techniques, is more frequent in qualitative approaches than in quantitative ones, possibly because spatial covariation between taxa or biotas in accordance to a general causative model is not always apparent. In vicariance biogeography one often operates qualitatively and weights according to geological criteria. This prevents interpretation in terms other than geology, a practice that is commoner in qualitative approaches than in quantitative ones, thereby forming a drawback in the former.

The remainder of Part I concentrates on the explicit quantitative approach.

3

Methodology of quantitative biogeographical classification

This chapter starts with a model basic to most biogeographical classifications. From this, several classification criteria can be derived, which, in turn, can be formalized in several classification procedures. These procedures are known as similarity coefficients and cluster or partitioning techniques. Throughout I use the terms classification and cluster analysis interchangeably, although recently classification has included both cluster analysis and data assessment procedures.

A model of biogeographical classification

The centre of Figure 3 contains an extension of Schenck and Keen's (1936) model of biogeographical units. The arrows point to five criteria derived from it, each of which characterizes a particular parameter of the model and thus defines biogeographical units accordingly.

The model assumes that the taxa are restricted to a single unit only, and cluster in its centre. Since this model is based on taxonomic endemism, many resulting classifications are monothetic, being defined by presence or absence of particular taxa. However, the same criteria can be used in polythetic classifications that are based on characteristic combinations of taxa. Dropping endemism as a basic assumption does not affect the possibility of classifying biotas, although it may affect the dominance of historical interpretations of classifications. This interpretation is connected with assumptions such as a taxon's centrifugal dispersal, its historically increasing ecological dominance, or its permanent occurrence in refugia after major ecological disturbances such as the ice ages.

Criteria of biogeographical classification

The first classification criterion (Figure 4a; number 1 in Figure 3) assumes that the ranges of taxa coincide optimally within biogeographical

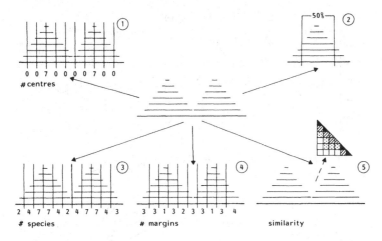

Figure 3. Model of biogeographical classificatory units with various classification criteria. For explanation, see the text.

units, the range centres lying closely together. Applying this criterion involves counting the range centres in sections of a linear transect or along a coast, or in quadrats of a grid partitioning an area. The section or quadrat with most range centres locates the centre of the biogeographical unit, whereas boundaries have the smallest numbers.

The second criterion characterizes biogeographical units by the number of taxa endemic to certain areas (Figure 4b). Its procedure consists of counting the most restricted taxa in the linear sections or squares. The unit's centre has most of the restricted taxa, and the borders between them the least.

Third, biogeographical units can also be characterized by a relatively large number of taxa (Figure 4c), and the boundary areas by relatively low values. Here, one counts the species present in each section or square for comparing their numbers.

Contrary to these criteria, the fourth criterion emphasizes boundaries between the units (Figure 4d). These can be found in that part of the transect or grid with most range margins. Here one counts the range margins in various parts of the transect or grid to compare these numbers (see Dale (1988) for a test of clumping of range limits along gradients).

The fifth criterion emphasizes qualitative compositions of biogeographical units. To discriminate between units, one estimates the number of taxa specific to sections of the transect or squares relative to

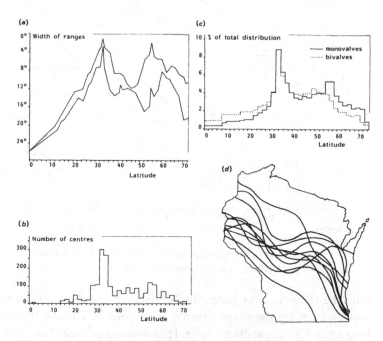

Figure 4. Application of four of the classification criteria in Figure 3. Figures *a*, *b* and *c* pertain to mollusc data in Schenck and Keen (1936), and Figure 4*d* to plant data in Curtis (1959).

that shared. Most (dis)similarity coefficients are procedures based on this criterion.

A more recently developed sixth criterion, not indicated in Figure 3, assumes that biogeographical units are relatively homogeneous, whereas pooled collections of taxa from different areas are more heterogeneous. Homogeneity can be estimated using sample variance or diversity coefficients developed in information theory. The value of these coefficients is low when the pooled collections are homogeneous and high when they are heterogeneous. Using divisive partitioning techniques, one first estimates the degree of homogeneity of the total area, then that of the total area minus one of the subareas, and so on; dividing lines are drawn where the variance or diversity changes most (e.g. Pielou, 1979*b*).

Diversity can be calculated using several formulae, of which Shannon and Weaver's (1949) coefficient is most commonly used (e.g. Järvinen and Vaisänen, 1980; Kikkawa and Pearse, 1969). However, these measures have no biological interpretation, for, contrary to the other five criteria, they do not retain topographical information of distribution ranges. Consequently, areas may be pooled because of a similar homogeneity, although they contain different taxa. For example, Ezcurra, Rapoport and Marino (1978) pooled the tundra and the Sahara together because of their relatively low diversity, which was obviously due to their small numbers of taxa, rather than compositional similarity! Moreover, the values that diversity coefficients take depend on the spatial dispersion of the individuals or taxa used, which may vary (e.g. Hengeveld, Becker and van Biezen, 1982).

Biotas can also be classified by using a combination of several criteria. Van den Hoek (1975), for example, estimated the number of species of green algae endemic to several parts of the North Atlantic (criterion 2), and the decrease in this number between those parts (criterion 3). Hagmeier and Stults (1964) and Hagmeier (1966) first estimated 'primary areas' using the geographical variation in the number of range margins over North America (criterion 4), and then estimated their compositional resemblance (criterion 5).

The fact that several criteria characterize various properties of one classification model makes their combined use preferable to using them separately. However, one may weight them unconsciously when choosing between them for the comparison of samples, as, for example, Van den Hoek (1975) did. This introduces a non-explicit element into the application of explicit procedures. Also, different sets of criteria may give different results (cf. Kaiser, Lefkovitch and Howden, 1972). Choosing the criterion that fits the purpose of the classification closest makes interpreting the results easier than choosing a combination of several of them.

Problems in the application of classification criteria
Endemism as a classification criterion
In monothetic classifications and according to some classification criteria, biogeographical units are distinguished by the number of endemics. This has even resulted in adopting local endemism in formal definitions of such units (e.g. Ekman, 1953). To judge the correctness of this criterion, we must know the actual and expected geographical patterns of endemics.

If endemic taxa characterize biogeographical units, they should occur at all levels of the classificatory hierarchy and homogeneously within each of them. Lowest levels would then be characterized by the most restricted taxa and the highest levels by the most widespread. Moreover, endemics should be distributed evenly over the total area and not be restricted to certain parts only. Information seems to be available only on the second aspect.

Clayton and co-workers calculated similarities of grass floras in several parts of the world to construct minimum-spanning trees, for practical reasons excluding widely occurring species and endemics from the actual calculations (e.g. Clayton and Panigrahi, 1974). After classifying the area concerned, the percentage of endemics was calculated for each unit distinguished. It then appeared that the geographical distribution of endemic grass species varies considerably, some regions being almost devoid of them, whereas others are centres of endemism. In Africa, for example, endemics particularly occur in Ethiopia and East Africa (Clayton, 1976) and hardly at all in other parts. India, Deccan and the Himalayas are centres of endemism (Clayton and Panigrahi, 1974) and the Middle East, including India, has several such centres. At this level, endemic species are thus unevenly distributed, occurring in various proportions in the clusters. Endemics do not thus always characterize biogeographical units and, consequently, cannot serve as classification criteria or to define them. Biogeographical units should be defined polythetically.

Effects of defining sampling areas

Partitioning areas will also affect classifications of the subareas, different sizes of subareas giving different results. Phipps (1975) developed a technique for optimizing this size for particular scales and taxa. This technique obviously cannot be applied to areas where topographical aspects interfere, for example with islands.

Another effect results from dividing continuous spatial trends. When continuous trends in biotic composition occur in all directions, similarity will be greatest in the centre of the area chosen and fall off towards its margin, subareas at opposite ends of this area having less taxa in common than those nearer by (Figure 5). This effect is strongest for low-level taxa, such as species, and weakest for high-level taxa, such as families. In general it is also stronger in large areas, such as the Pacific or continents, than in smaller areas. Clusters can thus originate at the margins of large geographical areas, especially if sampling is heterogeneous, or when islands are widely spaced, separated by different sized stretches of water. Such clusters are obviously artificial.

The spatial orientation of sampling transects relative to trends in biotic

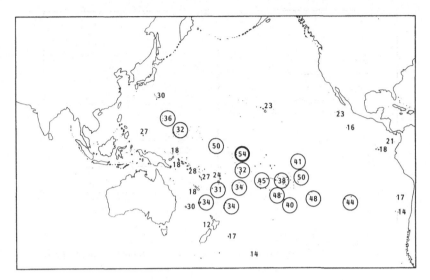

Figure 5. Average similarity of the plant genera found on 36 Pacific islands (after data in Van Balgooy, 1971).

composition also gives problems in classification, particularly for criterion 4. If, for example, a transect runs perpendicularly to such a trend, the number of margins per section of this transect is highest, but it decreases the smaller the angle, eventually being zero when the transect runs parallel to the trend. Other problems arise when part of the margins parallel each other, whereas other parts follow other directions. Curtis (1959) therefore excluded some species from his analysis of plants of Wisconsin. Depending on the proportion of excluded taxa and on the problem to be solved, one runs the risk of applying potentially subjective criteria and thus of introducing circular reasoning.

It is usually difficult to delimit sampling areas, which has often been avoided by adopting a geomorphological or ecological discontinuity. Apart from methodological problems that arise from adopting extrinsic discontinuities, the danger exists of generating artificial boundaries. Figure 6 shows such a discontinuity in a continuous trend in range locations generated by arbitrarily dividing it into two. The Jaccard similarity in this case is 0·28. Although, in principle, this difficulty can be avoided, it remains where biotic trends are divided naturally, for example in archipelagos where biotas are separated by the sea. Here clear classifications can arise, reflecting the size of the physical discontinuities.

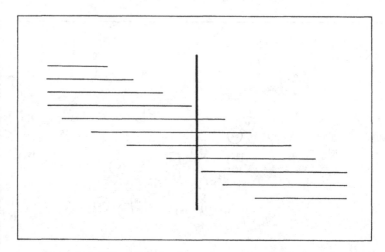

Figure 6. Arbitrary division of a continuous, hypothetical trend in species occurrences along a gradient.

Apart from selecting locations of dividing lines in a continuous trend, the steepness of this trend and the size of the ranges and of the sampling areas relative to the area the trend covers can also determine similarity values. As changes in location, shape, size and orientation of ranges may not always be in parallel in different species, geographical patterns of numbers of range margins may also alter. Thus the use of the number of range margins as a criterion of classification (number 4) will give less stable results than that of biotic composition.

Similar difficulties occur in the temporal delimitation of the period a particular classification should apply. When classifying biotas, one assumes that the ranges of most taxa remain more or less stable, since any change in the location or shape of distribution ranges will affect the biotic composition of the areas compared, and, consequently, the resulting classification. Biogeographical classifications will vary from one period to another, historical ones differing to a greater or lesser extent from present-day classifications (e.g. Birks, Deacon and Peglar, 1975; Bush, 1988). Since ranges of higher taxonomic levels are often more stable than those of taxa at lower levels, we can choose longer periods for the former than for the latter. The length of period chosen will depend on the taxa, because of differences in range dynamics between them, as well as on the size of the total area and subareas. For example, species composition of the carabid fauna in a relatively small area as the Netherlands has greatly varied over the last nine decades (Hengeveld, 1985*b*), although it may have been more stable on a European scale.

Procedures in biogeographical classification

After discussing a model of biogeographical classification and classification criteria derived from it, I consider similarity coefficients and cluster techniques as classification procedures. I confine my discussion to emphasize the heuristic nature of the results and to the possibility of testing discontinuities (cf. Birks, 1987; Gordon, 1981).

Similarity coefficients

A plethora of measures has been developed for expressing the biotic similarity or dissimilarity between two areas (Cheetham and Hazel, 1969; Sneath and Sokal, 1973). Cheetham and Hazel (1969) showed that many of them converge to Simpson's (1960) coefficient when the numbers of taxa that two areas do not share are identical. Apart from these, others express the relative degree of overlap of distribution ranges, and still others, particularly diversity coefficients, the relative degree of homogeneity of biotas.

Simpson (1960) developed his coefficient $S=c/a$ (where c is the number of taxa two biotas share, and a that in the smaller biota; Figure 7) for cases where sampling is inadequate or where the smaller sample is an impoverished derivative of the larger. One disadvantage is that it stresses similarity, sometimes even hiding structure in the data, preventing endemic and cosmopolitan biotas to be discriminated (Rowell, McBride and Palmer, 1973). Another disadvantage is that the results are unstable for low values of a, small changes having disproportionately large effects (Hughes, 1973). Conversely, dissimilarity can be emphasized, as in Braun-Blanquet's (1932) coefficient $S=c/b$, where b is the number of taxa in the larger biota. This coefficient is more stable as b represents relatively large numbers. Simpson's coefficient can be used when interest is centred on migration and Braun-Blanquet's when endemism, biological impoverishment, or barriers to migration are to be stressed, the latter emphasizing historical aspects of geographical distribution. Simpson's coefficient emphasizes similarity, and Braun-Blanquet's dissimilarity. Other coefficients weight a and b differently relative to each other or to c.

Instead of similarity, we can estimate dissimilarity, which are often

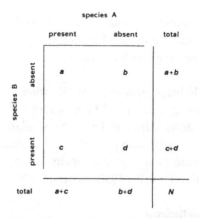

Figure 7. Definition of parameters used in (dis)similarity coefficients.

related as $D=1-S$. As they can give identical results, similarity coefficients do not necessarily emphasize equality and dissimilarity coefficients inequality. Dissimilarity coefficients, such as the Manhattan or city block distance, Euclidean distance, and Mahalanobis's generalized distance D^2 often relate to similarity coefficients (Gower, 1967). Mahalanobis's D^2 has the advantage that it allows for correlations between variables, which other measures do not. Preston's (1962) dissimilarity coefficient, $x^{\frac{1}{2}}+y^{\frac{1}{2}}=1$ in which x and y are the fractions of the total biota represented in the sample and z expresses dissimilarity, although it is known as the 'resemblance equation'. The values of x and y can be calculated according to $x=a/(a+b-c)$ and $y=b/(a+b-c)$, or as $(a+c)^n/(a+b+c) + (a+b)^n/(a+b+c) = 1$, where a, b and c are defined as above. Its disadvantage is that its calculation is complicated; z can only be approximated. To facilitate this, Preston (1962) gives a table for approximate values of z when x and y have been calculated. For $x > y$, z can be calculated as $z = -3{\cdot}32 \log(0{\cdot}6x + 0{\cdot}4y)$. When a and b are identical, the resemblance equation has no solution (Cheetham and Hazel, 1969).

Apart from these difficulties, Preston's coefficient is also controversial. Clifford and Stephenson (1975) and Simpson (1980), for example, argue that its complexity does not justify its use with binary data. Williamson (1981), however, prefers it as it would be mathematically derived from (1) the logarithmic relationship between species number and area, and (2) Preston's (1948) log-normal abundance distribution. I think it is dubious whether it can be log-normal, being based on a regrouping of abundance categories into geometrical classes, rather than

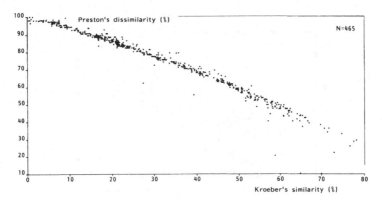

Figure 8. Plot of Kroeber's similarity coefficients against Preston's dissimilarity coefficient (i.e. resemblance equation) for Pacific plant genera (after data in Van Balgooy, 1971).

on logarithmic transformation, required for log-normal distributions. Depending on the steepness of the distribution of the raw data and the definition of the geometrical classes, any curve may result, be it down-sloping, humped, uniform, or up-sloping. A better description of the species-abundance distribution is Fisher's (in Fisher, Corbet and Williams, 1943) log-series distribution, as an approximation to the general models of Hengeveld and Stam (1978). Moreover, the parameter z, expressing dissimilarity is equal to z in $s = ka^*z$, which describes the relationship between species number (s) and area (a), although the value of z is a mathematical artifact (May, 1975; cf. also Connor and McCoy, 1979).

Finally, the advantage of using Preston's coefficient does not appear from the high correlation between 465 dissimilarity values according to Preston (1962) and similarity values according to Kroeber (1916) for Van Balgooy's (1971) Pacific plant genera (Figure 8).

Effects of the number of shared taxa and of sampling area

Many similarity coefficients contain parameter c, as in, for example, Kroeber's (1916) similarity coefficient $S = c(a+b)/2ab$. Correlating c with these similarities, c appears to account for half ($r^2 = 0.51$; $n = 630$) of the total variance in the data (Figure 9a). This amount of variance explained by c may vary for other data and other coefficients, resulting in classifications being partly explained by factors that make biotas different and partly similar.

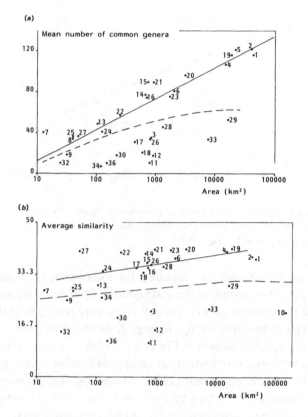

Figure 9. Number of common genera (*a*) and average similarity (*b*) between 36 Pacific islands as a function of surface area of these islands. The dots below the dashed line represent islands that are remote, depauperate, or insufficiently known (after data in Van Balgooy, 1971).

Although similarity values partly depend on the number of shared taxa, this number, in its turn, depends on the relative sizes of the areas (Figure 9*b*). This relationship is positive because the number of shared genera is a fraction of the total number of genera present on an island, which relates logarithmically to island area. Since similarity is a function of the number of shared genera, it also relates to island area. Preston's (1962) resemblance equation would, according to Williamson (1981), compensate for differences in island (sample) size, which does not apply because of its correlation with Kroeber's (1916) similarities relating to *c* (Figure 8). However, despite its significance and although it can confuse interpretation, variation in sample size is usually not considered in biological classification.

Testing differences between similarity coefficients

Similarities have been developed to estimate differences in multivariate data, such as biotic composition between areas, without basing these coefficients on theory. Their formulae are empirical, as is the criterion of efficiency; results are judged 'good' or 'bad', according to preconceptions, which include circular arguments. No independent alternative explanation or null-hypothesis is formulated, leaving one without a yardstick to estimate the probability that a similarity obtained may have biological meaning.

During recent years procedures have been developed to generate empirical distributions on which chance distributions are based (e.g. Birks, 1987; Henderson and Heron, 1977; Raup and Crick, 1979). The observed taxa are allocated at random to the areas many times, and each time similarities are computed from which a frequency distribution of similarities is constructed for all area comparisons. Because these distributions are constructed from random allocations of taxa, they serve as a yardstick for the likelihood that actual observations deviate from random expectation. If significant deviations exist, we can infer that the patterns result from biological or historical processes.

Raup and Crick (1979) give three examples of applying this procedure, one of which concerns global distributions of 222 echinoderm genera. They showed that similarities are highest close to an arbitrary reference point near Central America, these values decreasing with distance and becoming significant. Mapping probability levels of the similarity values thus adds extra information useful for interpretation. Rice and Belland (1982) recently applied a related procedure to construct empirical probability distributions for Jaccard similarity values.

In this case all known echinoderm genera ($N=a+b+c+d$ in Figure 7) were used, which is usually difficult in biogeography, if not impossible, as N or d depend on an arbitrary definition of the source areas for the taxa or the region under study (cf. McCoy and Heck, 1987). This arbitrariness gives statistical testing a subjective basis, although its purpose is to enhance objectivity. Reasons for including d are that existing theoretical probability distributions can be used to test the significance of the observations (e.g. Goodall, 1964, 1966).

In general, when treated in combination, statistically non-significant values may show up significant patterns. This treatment belongs to the domains of cluster analysis and ordination discussed later.

Recommendations

Many properties of similarity and dissimilarity coefficients are insufficiently known, such as effects of shared taxa, size of sampling area, and biotic diversity. Williams (1947) first corrected for differences in area by assuming Fisher's (in Fisher *et al.*, 1943) log-series distribution and only after that for diversity variation, measured by parameter α (Williams, 1949). He thus reformulated Jaccard's similarity in terms of diversity and number of species in a sample. Jaccard's coefficient appears to rise with increasing sample size, and to fall with higher species numbers. Wolda (1981), considering 100 000 hypothetical individuals, distributed among species according to the log-series distribution, compared many coefficients by taking sample size and diversity into account. He recommended Horn's (1966) simplified version of Morishita's coefficient, which, however, requires N to be known.

For presence–absence data, Euclidean distance and Jaccard's coefficient give the fewest problems, and most is known about their properties. Their equations are simple and easy to compute, and, most important, the results are easy to interpret. Jaccard's coefficient can also be rewritten in terms of log-series distributions from which its dependence on sample size, species number, and diversity can be seen. By knowing these relationships one can correct for variation in these parameters on the similarity value.

Thus, these two coefficients are the most flexible ones presently available for presence–absence (binary) data commonly in use in biogeographical classification. As they are also the ones most widely used, their application is recommended.

Hierarchical cluster techniques

This section surveys several cluster techniques, showing differences in partitioning criteria they adopt, rather than mathematical detail (cf. Cormack, 1971; Gordon, 1981; Pielou, 1984). Although these criteria are all reasonable, their results often differ, making the selection of classificatory procedure, and hence the results obtained, arbitrary. As this is a limitation, rather than a criticism of the procedures, classifications remain useful in exploratory phases of research.

Cluster algorithms can be distinguished as divisive and agglomerative. Divisive algorithms divide the original data set, i.e. areas or taxa, into several subsets according to prescribed criteria. This is done step-wise; first the set is divided into two, then each of these two again into two, and so on. Some measure of variation, taken as the partitioning criterion, is minimized at each step to define the subsets. Kikkawa and Pearse (1969), for example, used a diversity index as a criterion for partitioning the Australian avifauna.

Agglomerative clusterings employ the opposite strategy: instead of considering the entire data set as the first cluster, these techniques define each individual as a separate cluster. Then, step by step, mutual similarities between the initial clusters are sorted, so that they gradually form clusters at higher levels, until all individuals are united into one single cluster equal to the total data set.

A generalized algorithm for agglomerative strategies

Lance and Williams (1966, 1967) and Wishart (1969) developed a general clustering algorithm in which various parameter combinations lead to the algorithm for single-linkage, complete-linkage, centroid, and median grouping, weighted and unweighted group-average, and Ward's sum-of-squares methods. Lance and Williams added another technique, flexible sorting, derived from this generalized algorithm. These cluster techniques are thus particular forms of one general solution with constraints defined by the value that each of three parameters α, β and γ take in the equation

$$d_{hk} = \alpha_i d_{hi} + \alpha_j d_{hj} + \beta d_{ij} + \gamma |d_{hi} - d_{hi}|.$$

Here, d_{hi}, d_{hj}, and d_{ij} are the distances between three groups of individuals in multidimensional space (Figure 10) and d_{ij} is the distance between one (h) of these three groups, with the fused other two groups, i and j. Parameters α, β, and γ define the cluster strategies, their values differing between them. Sometimes, as in single-linkage, complete-linkage, and in median grouping, they are constants, and in other cases one or more of them depend on the number of individuals n_i, n_j, and n_h in the three groups i, j and h, respectively. In group average and centroid clustering α_i and α_j are defined by the proportion of individuals group i and j contribute to their combination, group k. The parameter β, linked with the distance between groups i and j, is either constant, or depends on the number of individuals in the various groups. It affects distances between groups i and j relative to that of either of these two groups relative to h, such that few distinct clusters are found when β is large and positive, and many,

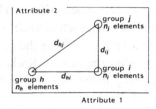

Figure 10. Relationships between three groups of individuals (biotas) relative to two attributes (taxa).

very distinct clusters when β is large and negative. Positive values of β are space-contracting (that is, it brings group h relatively close to group k), a zero value space-conserving (leaving the original relative distances unaffected), and negative values space-dilating (enlarging distances d_{hi} and d_{hj} relative to d_{ij}). As weighting the distance between groups relative to each other influences the ease with which fusion occurs, β is called the 'cluster intensity coefficient'. It may or may not depend on the number of individuals within each group. Parameter γ usually adopts a zero value; only in single- and complete-linkage does it weight distances between group h and groups i and j by a constant, -0.5 and $+0.5$, respectively.

Lance and Williams (1967) showed that β can be given any value in addition to those in existing strategies. As β can assume any value between -1.00 and $+1.00$, cluster intensity is a continuous function and makes clustering flexible. The values that characterize various strategies are thus arbitrary choices from a continuum of all possible values of β. Lance and Williams (1967) advised fixing β at the value of -0.25 in 'flexible' sorting, thereby dropping its flexibility! (See, for example, Clifford and Stephenson, 1975, for its use.)

As each cluster strategy assumes particular values for α_i, α_j, β, γ and n_k, cluster intensity varies. Intensity is defined by one or more constants, or by the number of individuals in the groups. Thus, the degree of data clustering is not only an inherent property of the data themselves, but also depends on an arbitrary choice of parameter values defining the selection of cluster technique and of sample size. These features are thus criteria for judging the clustering of a given data set. The next section gives a geometrically more tangible picture of the algorithms mentioned.

Some specific agglomerative cluster techniques

In single-linkage or nearest-neighbour strategy the clustering criterion is the smallest mutual distance between clusters. First, one finds

Methodology of quantitative biogeographical classification

those two individuals lying closest together; then those lying closest but one, and so on. Its limitation is that the smallest distance between clusters must be greater than the smallest distance between individuals in either of them; as many natural groupings overlap, this procedure often results in chaining. A minimum-spanning tree is a geometrical non-hierarchical variant of single-linkage clustering.

Complete-linkage or furthest-neighbour strategy avoids chaining by sorting individuals using their largest distances rather than the smallest ones. Consequently, distances between clusters are the distances between their most distant members; overlap between clusters thus becomes maximal. As groups grow, distances between groups increase, soon filling the whole multidimensional space. Although we can distinguish separate clusters, the strategy is not satisfactory.

The centroid strategy defines clusters by their mean locations in multidimensional (n individuals $\times$ m variables) space. As it is not geometrically a group average, this mean location is usually called the cluster's centroid or centre of gravity. After each fusion of an individual to existing clusters, the centroid is calculated anew. This procedure has the advantage over single-linkage and complete-linkage that it allows some, but not too much overlap between clusters; its drawback is that cluster shapes are assumed to be radially symmetric. If they are skewed, differences in the number of individuals in two clusters becomes critical, affecting distances between clusters. This strategy is still unsatisfactory, although it takes account of the locations of all members in a cluster to define differences between clusters.

The weighted centroid or median strategy circumvents effects of differences in the number of individuals by determining the n-dimensional median to characterize the location of all cluster members. This weighted centroid method weights the individuals differently, whereas in the former procedures they had equal weights. 'Weighted' therefore refers to the individuals themselves and not to the weight of their numbers as in the centroid strategy. Its drawback is that differences in compactness among clusters can still exist.

Contrary to the previous strategies, Ward's sum-of-squares strategy estimates differences in density of individuals in multidimensional space, defining areas of more or less homogeneous density as clusters. Squared Euclidean distances are used, which are then summed; these sums-of-squares are minimized for cluster recognition. Its limitation is that dissimilarities represent Euclidean distances; it should not be used with other coefficients.

Recommendations

Ward's method is generally the most useful, although Webster (1977) notes that once two individuals or groups are joined, they remain fused, however inappropriate their fusion might be at higher levels of classification, which, obviously, is true for all clusterings. Moreover, of the strategies discussed, it comes closest to general multivariate statistical methodology.

TWINSPAN, a divisive technique

Divisive clustering techniques divide the total data set up into two or more groups and repeat this for each of these two groups, and so on up to the point of reaching the individual sampling areas. Geometrically, it is difficult to divide a multidimensional scatter of points up in a hierarchical way, unless one knows the direction of the largest variation, that of the largest-but-one, and so on. TWINSPAN (Hill, 1979) therefore estimates these directions of variation by projecting axes in those directions. Then the points along each of these axes are divided into two relative to the dividing point chosen on the respective axes. The first axis describes the direction of greatest variation and hence the highest hierarchical level in the classification and subsequent axes gradually lower levels.

The consequence of dividing the total data set making several independent steps is that the hierarchical levels cannot be compared, neither horizontally, nor vertically. Their interpretation may differ in terms of the scale of spatial variation. As there is no reason to suppose that processes within either of two clusters are similar, the best thing one can do, therefore, is to treat them as being independent.

Problems in the application of clustering algorithms

Area delimitation

Methodology of biogeographical classification is complicated because the units to be classified cannot be delimited uniquely, only whole continents are most clearcut. Even then difficulties may arise, such as in the Palaearctic and Nearctic, which can be combined as the Holarctic. As no area is homogeneous (otherwise classifications would not make sense), altering the size of the area to be classified will affect the classification; classifications of the same general area, but with dissimilar limits, will not be comparable. Moreover, vast areas are required for taxa with large ranges, such as those of higher taxonomic levels or with great dispersal power, and either small or large areas for those with small ranges. The results of classifying one particular area for taxa with large or small ranges will also differ, as will those that either exclude or weight taxa with large ranges, the 'wides', or with small ranges, the endemics.

One may therefore choose a supposed barrier, some natural delimitation of the area, or an area of a certain degree of homogeneity, but, in so doing, one runs the risk of introducing a factor potentially explanatory for the classification at hand. One has to admit that classifications are heuristic and that they should be used only as starting points for further research, rather than as descriptions of objective, natural units, being ends in themselves.

Interlocation variation

Inhomogeneity of sampling may add to the discontinuity in biotic composition of an area, sometimes even generating it; when regular compositional clines are unevenly sampled, classifications may reflect the unevenness of sampling relative to the steepness of the cline. Moreover, when sampling is at regular intervals, sampling density may generate discontinuity of variation rather than the actual continuous variation (Figure 6). When several biogeographical units are distinguished, one assumes that the samples taken are representative, hence that effects of sampling error are relatively small; most taxa should have been treated, and excluding them should not bias the classification. Apart from observational errors, others result from chance processes; after dying out in some areas or islands, species may not have had the opportunity to recolonize their former locations. Chance processes like these should not bias classifications of an area. It is even more difficult, if at all possible, to discover effects of discordant variation.

Discordant variation describes spatially independent variation of character states of organisms, species, or higher-level taxa in taxonomy. For taxa to be recognized, it is necessary that many characters show the same geographical distribution, rather than independent distributions. If the characters covary in space, one is justified in distinguishing one or more taxa, otherwise one is not. High spatial covariation is called concordant character variation, and low spatial covariation discordant variation. Two measures of discordance are the cophenetic correlation coefficient (cf. Sneath and Sokal, 1973) and the coefficient of concordance (e.g. Jardine and Edmonds, 1974). As the term cophenetic correlation coefficient suggests, one estimates how far several characters vary similarly. Classifications very much depend on the discordance of the characters one happens to have chosen, often making it arbitrary and unstable; for high levels of discordance any newly added character will alter the classification. Distribution patterns of taxa within higher taxonomic levels may or may not greatly overlap and their classifications will, consequently, be arbitrary and unstable, depending on their biogeographical discordance.

When discordance is high, differences between several locations do not give interpretable subdivisions of a biota. In such cases, all taxa are distributed according to their individual responses to the same or sometimes even to different environmental factors, rather than to the same factors to which they all respond similarly.

Intralocation variation

Intralocation variation in biotic composition may be due to three factors: (1) differences in sampling coverage, (2) random extinction of taxa, and (3) size differences of the sampling location or of habitats within them.

Differences in sampling coverage may be quite sizeable, considering that Connor and Simberloff (1978) could relate the number of taxa known from individual Galapagos islands to the number of visits made to them! Moreover, in practice, sampling coverage is highest in well-delimited areas such as islands or lakes than in a certain, vaguely delimited part of a continent. Islands are often revisited when coverage seems incomplete, but some part of a continent is not revisited. It would therefore be interesting to allow for, for example, between-island variation within archipelagos before comparing several archipelagos among each other, so as to get an impression of discordant variation due to incomplete sampling.

Species may also for some reason or another become extinct on one island and not on another. Some islands may be wetter than others, be higher or lower, larger or smaller, volcanic or coral, and so on. Even slight changes in general ecological conditions may affect populations on various islands differently. Of course, one can add all species per archipelago together, but this leaves size variation of archipelagos due to differences in numbers of islands, and size variation among individual islands. The same holds, *a fortiori*, for areas on continents. Therefore, due to independent species behaviour, or of taxa in general, intralocation variation may be discordant.

Sampling locations may also differ in size, which, unless one corrects for size, makes them only partially comparable. Taking islands of several archipelagos again as an example, size distribution of islands within these archipelagos will usually vary. This too adds to discordance in biotic variation in geographical space.

Caution should therefore be taken when interpreting results of a classification; without these various effects being known, classifications are easily misinterpreted. Moreover, we should be cautious in proposing general classifications and in giving interpretations pertaining to the possible causation of distribution patterns of many taxa, or in generalizing to those of taxa not treated. In erecting classifications for particular, restricted taxa only, one attempts to increase the concordance of the data, which obviously introduces an element of circularity, preventing interpretation of the classification.

Testing classifications

High and low similarity values may be scattered randomly within a trellis-diagram, thus reflecting absence of spatial pattern in the composition of an area. If, on the other hand, all high values are concentrated along the diagonal and low values close to the corner, the composition changes gradually. In cases of discontinuous trends, high values occur in clumps near the diagonal and low values near the corner. Pielou's (1979*a*) approach for testing if biogeographical discontinuities exist involves testing whether clumping along the diagonal is significant.

However, her test statistic Q/Q-max cannot be used as it expresses the degree of concentration of values along the diagonal rather than clumping. Q/Q-max, namely, does not change when the clusters are defined differently and the sequence of individuals is kept the same, nor when, alternatively, the clusters are kept the same but when their sequence is altered.

Another approach to testing biogeographical classifications is to randomize the elements in the data matrix many times and to calculate each time clustering statistics using the same clustering strategy. The frequency distribution of all outcomes represents a random or null-frequency distribution. The actual dispersion of taxa is compared with this null distribution to test if it occupies an exceptional position relative to this distribution, occurring in one of its tails, or if it falls within the bulk of all outcomes.

Two problems arise when applying this approach: (1) the choice of test statistic representing the results of clustering; and (2) whether the marginal totals of both rows and columns of the data matrix should be held constant when randomizing this matrix, or that the row totals (i.e.

the number of positive values for each individual, either taxa or areas) should be held constant whereas column totals are allowed to vary. In the first problem, we can choose the nodal values in the dendrogram as a test statistic (Harper, 1978), and construct their frequency distribution for the first, second, third, . . . order of nodes (Strauss, 1982). From the null distributions of these nodal values, one establishes empirical 95% confidence limits for testing the deviation of observed values.

Contrary to the first problem, which is purely statistical, that concerning the constancy of the marginal totals is biological. Allowing these totals to vary in a classification of areas implies that numbers of taxa in various localities can differ from those actually observed, affecting assumptions concerning niche saturation and environmental carrying capacity. Harper (1978) did not fix marginal totals, whereas Strauss (1982) did. When one does not fix marginal totals, one can adopt various probability distributions for the spatial dispersions, being uniform, binomial, Poisson, or a mixture of Poisson distributions, depending on the process assumed to generate the dispersions.

Yet another approach to testing classifications concerns distributions of taxa along linear gradients, such as sea coasts or transects. This approach was originally formulated statistically by Pielou (1975) and reformulated later by Underwood (1978) (see also McCoy, Bell and Walters, 1986). It uses the number of range limits in each spatial unit (criterion 4) and thus differs from other approaches that use biotic composition as a classification criterion (5). It follows a theoretical probability distribution instead of an empirical null-frequency distribution. Classifications based on diversity indices (criterion 6), however, cannot be tested.

Conclusions

I presented a general model of biogeographical classification to provide a general background. Though based on endemism, occurrence of endemics is not essential. From this model several classification criteria were derived, each characterizing the model's spatial unit in a particular way. In the resemblance criteria, various calculation procedures and similarity coefficients can be formulated. Similarities form the starting point for calculating clusters. Again many algorithms exist, all giving different results. The statistical significance of similarities and cluster distinction can be evaluated.

The quantitative biogeographer thus has great freedom to choose what seems the 'best' technique. Although the methods make classificatory procedures more explicit, the choice between various techniques is largely subjective. By admitting that this is so, we have progressed relative to the intuitive approach, because others now know exactly what choice was made and what its consequences are for interpreting the results. In disagreement, they can improve upon this following their own ideas, and hence advance our science.

This unavoidable subjectivity may seem a sad state of affairs for any science, especially when we realize that the results from these choices can vary markedly. Results will also deviate for data sets differing in (1) the taxa used, due to their ecology, motility, and history, (2) the location and delimitation of the areas investigated, and (3) the subareas within them. But for a fair judgement, we must emphasize that these classificatory techniques are used to detect patterns and hence to generate hypotheses about possible processes. As long as these processes are unknown, we must make non-explicit assumptions and choices and evaluate the results. In this vein it is essential to treat results heuristically and to consider them as a starting point for further research involving testing of hypotheses about ecological and historical processes on a geographical scale. Classifications are not ends in themselves but means to an end.

4

Criticism of biogeographical classification

Since the 1950s, various aspects of biogeographical classification have been severely criticized. First, an important objection is that bio-geographical units imply a certain homogeneity, which may often be lacking. Second, classifications are unstable because of the dynamism of species ranges. Other criticisms dispute the nature of boundaries and emphasize the broadness of transitional areas between units and the impossibility of proving the physical reality of the units, and question the use of qualitative data. Also, classification is static and descriptive, whereas understanding biological processes on a geographical scale requires a dynamic, analytical approach. Practical utility is also questioned, as is the interpretation of classification, as it can overstress biotic dissimilarity rather than similarity. Drawbacks of qualitative approaches were mentioned above, along with certain technical and interpretative aspects of quantitative techniques.

Classifiers do not seem impressed by these objections and argue that classifications, however obtained, are often concordant, and contribute to our understanding of historical and ecological origins of biotas. This chapter evaluates some of the most important objections, starting from the idea that criticisms should improve future analyses, rather than be ignored or opposed. Incorporating criticism can stimulate research; otherwise it paralyses it.

Biogeographical classification and taxonomic level

When classifying, one problem is the choice of taxonomic level, as taxa vary at different spatial scales. Distribution ranges of genera will, at least, equal the largest range of its species, if all ranges coincide with it, if not exceed that of the largest species range. Similarly, family ranges are larger than those of their largest genera, and so on, implying that high taxonomic levels require investigation of large areas. As De Lattin (1957) felt that large areas would make the data heterogeneous, he suggested using species ranges only as these would be homogeneous, comparable, and taxonomically real units with a geologically similar age.

Others think otherwise. Raup and Crick (1979), for example, compared global distributions of echinoderm genera with those of species. Their results suggest that these distributions differ. For regions east and west of the Panama Isthmus the classification of echinoderm genera is explicable historically and of species ecologically. Similarly, classifications of distribution patterns of taxa at the level of Australian bird species, polytypic species, and genera can be compared, suggesting Pleistocene climatic changes have affected these birds, causing their restriction to isolated refugia. From here they spread, forming species adapted to new conditions (Kikkawa and Pearse, 1969).

It is thus not necessary to restrict analysis to one taxonomic level only; on the contrary, different levels may give interpretations of different biological relevance. Distribution patterns of higher taxa, possibly reflecting those of lower taxa in the geological past, may be compared with distribution patterns of lower taxa at present if this does not necessitate extending the analysis too far back into history. Comparisons may show geographical and evolutionary dynamics of taxa through time.

Are biogeographical units homogeneous and stable?

Some authors suggest that the homogeneity and stability of biogeographical units in space and time would be so great that they constitute independent entities (e.g. Briggs, 1974; MacArthur, 1972; Valentine, 1968). Their concordance would originate when species are highly interdependent. Under changing conditions the ranges of constituent species would therefore shift in unison, rather than independently in different directions and at various rates. However, other authors feel that biogeographical units have no physical reality; they cannot be observed and are not sharply delimited. They consist of historically and ecologically different elements, and are internally and externally dynamic.

Mayr (1965a), particularly, stresses their internal heterogeneity and instability and argues that they should be analysed internally, rather than classified as units. Apart from species' presence or absence, one should consider the degree of isolation of an area, and whether species have a different origin, adaptation, and age. Mayr distinguished six faunal types, originating from (1) autochthonous adaptive radiation of taxa; (2) continuous colonization from one, or (3) more areas; (4) fusion of two faunas; (5) specialized biotopes, such as mountains, deep seas, or deserts; and (6) a compound origin. Biotas should, accordingly, be analysed in a way that equally holds for the causative factors, such as climate, physiography, and specific geographical conditions.

Parts II and III discuss the complexity of spatial distributions of taxa, and the dynamics that Mayr may have meant. But classification remains the first stage in the description and analysis of biogeographical patterns, prior to the process analysis that Mayr proposes. The various nodes in dendrograms suggest the occurrence of processes at particular spatio-temporal scales. Conversely, using a particular spatio-temporal scale and taxonomic level implies that one analyses a certain process and not another. Biogeographical classifications reveal, heuristically, a spatial structure of partly contiguous areas with a certain biota. Given a taxon sampled at a certain spatio–temporal scale, and given a certain nodal value from a cluster analysis, the spatial units obtained are homogeneous and stable. But they are heterogeneous and unstable at another level of variation, with respect to another technique, another taxonomic level, and so on. They are also heterogeneous when properties other than range topography are chosen, such as growth form, morphological or ecological properties, or dispersal characteristics. One then uses different information which is usually not available when classifying the taxa on distributional criteria. Of course, we can add this information to that of distribution straightaway, but usually it is more profitable to do this during the hypothesis-testing phase of analysis. Thus, species within a classificatory unit conform in their distribution, but not necessarily in other respects as well, which also holds for Mayr's (1965*a*) categories based on other classification criteria.

As one's choice of classification criterion at the hypothesis-generation stage is open, different classifications do not necessarily contradict each other; they may parallel each other during investigation. However, Mayr (1965*a*) challenges the idea that species in biotas cannot interchange between these, and that biotas move as stable units when conditions change. Mayr thinks that biotic composition changes continuously, quantitatively and qualitatively, and that species intermingle freely, ever forming new combinations. There is much to say about this view and, as we will see later, much in favour of it.

Geographical discordance of ranges

Categories at a given hierarchical level are homogeneous when boundaries defined by a cluster analysis are applied to a data set, although going upward from lower levels, higher levels become heterogeneous, which is the same as enlarging the spatial scale of investigation to compare differences in taxa of the various biotas. Both approaches together are important in the analysis and interpretation of specific responses of the taxa concerned, differences in biotic content revealing specific responses to environmental conditions for each taxon individually. Lack of congruence in taxonomic content is another criticism of classification that can improve biogeographical methodology.

Species' ecological responses may vary according to life-cycle stage, adults having different ecological tolerances and requirements than larvae, or reproductive conditions may differ from those for adult survival (Hutchins, 1947). Coincidence of such conditions in one area does not mean that it will occur elsewhere as well. Each area is unique, which prevents detailed intercomparisons. As an example, Figure 11 shows variation in geographical limits of temperature fluctuation for the North American Atlantic coast. Species limited there by minimum and maximum summer temperatures of 7·2°C and 26.7°C (number 1 in Figure 11a) have the same extension as those limited by winter ice in the north and maximum winter temperature of 10°C in the south (number 4). However, their latitudinal extensions along the European, Asian, and western North American coasts differ. The southern limits of *Balanus balanoides* and *Mytilus edulis* coincide along the North American Atlantic coast, but not elsewhere, reproduction in *Balanus balanoides* presumably requiring low temperatures, which limits it in the south to a minimum winter monthly mean temperature of 7·2°C (Figure 11c). *Mytilus edulis*, in contrast, is limited in the south where maximum summer temperatures exceed 29°C, above which adults die (Figure 11b). As these temperatures are uncorrelated, distributional overlap varies along the different coasts.

The criticism of unique responses to unique combinations of environmental components is useful because regional discordance in biotic content reveals the impact of several causative factors, which sometimes overlap, but sometimes do not. Absence of such incongruences just limits the analysis.

How sharp are boundaries?

Boundaries between classificatory units are usually vague and constitute broad transition zones. This is because the main part of biotic ranges may coincide with particular units, although the taxa often occur in other units as well. When this happens, such taxa are called biogeo-

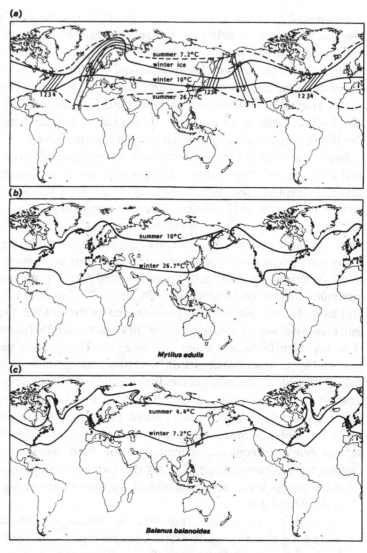

Figure 11. (*a*) Ranges of fluctuation between several characteristics of annual sea surface temperatures. Differences between (1) summer temperature of 26·7°C and winter ice; (2) summer temperature of 7·2°C and winter temperature of 10°C; (3) summer temperatures of 26·7°C and 7·2°; and (4) temperature of 10°C and ice cover in winter (after Hutchins, 1947). (*b*) Distribution ranges of *Mytilus edulis* explained by summer and winter sea surface temperatures (after Hutchins, 1947). (*c*) Distribution range of *Balanus balanoides* explained by summer and winter sea surface temperatures (after Hutchins, 1947).

graphic elements (Wulff, 1943). Polythetically defined units nearly always have vague boundaries, their component taxa declining gradually from their centre and merging into surrounding units.

Many factors determine the sharpness and location of boundaries between units; for example, taxonomic identity, taxonomic level, frequency distribution of range sizes of subtaxa within the taxon concerned, and number, scale, and delimitation of areas or subareas investigated. Moreover, arbitrariness in choice of the classification criterion and numerical procedure adds to the vagueness of the boundaries. Finally, as results from quantitative and qualitative data may differ (Lincoln, 1975), the reality of the boundaries is difficult to assess. Criticisms of biogeographical classification therefore often concern the 'reality' of boundaries and hence the classification defining them (e.g. Peters, 1955; Simpson, 1977).

However, in principle, boundaries need not be real and sharp. It is important to know which taxa contribute to their location and sharpness and which do not. Particularly their sharpness affects the nature and ease of further analytical research following classification. Broad transition zones can, in fact, assist in explaining the nature of the units, since more opportunities occur for testing alternative hypotheses. For example, to test the importance of ecological factors on distribution patterns in the Indo–Australian area relative to historical ones, Lincoln (1975) compared bird species in the transition between the Oriental and Australian faunas with regard to biotope preference and food preference (Figure 12). Particularly revealing, for example, were pockets of arid habitat on mountain slopes within a generally humid area.

Sharpness of boundaries is thus partly determined by the scale of the data, such as age, mobility of taxa (e.g. Holloway and Jardine, 1968), and taxonomic level, partly by sampling intensity, and partly by the numerical technique used. When boundaries are sharp, one might even alter the classificatory criteria to make them vaguer to facilitate analysis of the nature and causation of the units concerned.

Why are biogeographical classifications hierarchical?

In our classification model, distribution ranges of two sets of taxa are compared, using several criteria. Of these, the qualitative difference between two biotas (criterion 5) is applied most frequently, and the ecological or historical interpretations are inferred from a system of relationships that are hierarchically arranged. One may wonder if a hierarchical arrangement is essential for expressing relationships

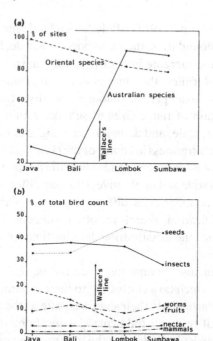

Figure 12. Transition between Oriental and Australian bird faunas along the
Lesser Sunda Islands. This transition as measured by the proportional
number of sites where these two groups of birds are found (*a*) and as
measured by dietary groups in which the total bird fauna was divided (*b*)
(after Lincoln, 1975).

between the units, or what the interpretation is of the hierarchical levels.
Why not use a non-hierarchic classification in some cases?

Partitioning areas into hierarchical systems may be done for practical
reasons or it may stem from taxonomic classification where it reflects
natural relationships of discrete taxa, having a common phylogeny, age,
and historical development. Sometimes, however, spatial variation may
be gradual, clinal, or discordant, with broad transitions and intergrada-
tions, showing no clusters at all. Hierarchical clustering describes relationships between discrete entities, whereas non-hierarchical clustering
considers entities with some overlap, except, for example, optimization
methods (e.g. Gordon, 1981).

Change of technique implies asking different questions, and results in
different perspectives. Hierarchical clusterings produce clusters with
discrete nodal levels, each with a different interpretation. Comparing

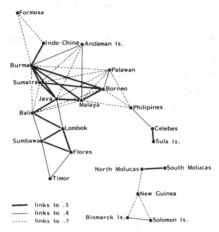

Figure 13. Faunal similarity of Malaysian islands at various similarity levels (after J.D. Holloway, personal communication).

clusters at the same hierarchical level is common practice in taxonomy where within certain taxa genera, families, or orders are studied in their own right. Hierarchical classifications, in which interest is centred on relationships between discrete entities differ from non-hierarchic ones that emphasize trends or transitions between entities, if the latter can be distinguished at all. In B-k or k-clustering (Jardine and Sibson, 1971), an arbitrary chosen similarity level allows for a certain overlap, where the parameter k, the number of individuals two clusters share, can be zero (B-1, or single-linkage clustering), one (B-2) (Figure 13), two (B-3), or three (B-4), etc. Chains of clusters can be graphically represented, showing clinal differentiation of a chosen degree (e.g. Jardine, 1971). Instead of the number of individuals as a criterion of overlap, we can also define overlap diameter (Jardine and Sibson, 1971), resulting in u-diametric clustering. Interpretation in these cases concerns causes of spatial trends and shows that clustering can merge into ordination.

Our interest here is whether biogeographical relationships within a hierarchy all have the same interpretation, or a different one. Do taxa in low-level units within a hierarchically higher unit have a similar ecology, phylogeny, or geographical history? Are higher units explained mainly historically and lower units ecologically? Should hierarchically low-level units be analysed differently than high-level ones, for example by non-hierarchical techniques rather than hierarchical ones? Is the structure of biogeographical hierarchical systems heterogeneous, higher levels containing the most widely occurring taxa, and lower levels the most restricted ones?

I feel that for each biogeographical classification of different taxa, hierarchical levels can have different interpretations, although it is unclear why information from those levels has been so little, if at all, used. As long as we do not know the ecological and historical causes of biogeographical patterns and processes, we must use information from all individual levels separately.

Conclusions

Biotic classification is one kind of biogeographical analysis and, within this, we can have very different aims. For example, constructing a reference system may be one; initiating a historical or ecological analysis may be another, which requires new observations to be made for testing them. We should not stop after drawing dividing lines resulting from a certain clustering; drawing such lines is the actual start of biogeographical research.

Essentially, the various problems in this chapter refer to an outdated problem, as polythetic classifications have now largely replaced monothetic ones. Usually biotic discontinuities indicate regions where trends in biotic change are steeper than elsewhere. These can be described by polythetic techniques and not by monothetic ones. Monothetic techniques are based on endemism and sharp, qualitative changes at, consequently, sharp boundaries. Ranges of constituent taxa are assumed to be concordant and stable, and the units homogeneous. Often taxa or regions are weighted individually. Trends or migration cannot therefore be handled by these qualitative techniques. In contrast, polythetic techniques are based on compositional differences as the classification criterion. Differences between units are those in degree and their boundaries are transitional zones. Ranges can be discordant to some degree and the units, accordingly, heterogeneous. As the units differ in degree, weighting can be done by the weight of numbers, rather than by individual properties. They allow some dynamics among taxa by migration, speciation, or extinction and as they are based on, for example, spatial variation in percentage change, some techniques describe trends, and thus merge with ordination. Much of the criticism of biogeographic classification concerns drawbacks inherent in outdated approaches, whereas some concerns misunderstanding of its aims. Formerly these aims were to produce regional classifications of general applicability, often lacking a detailed, tested interpretation. At present it aims at generating interpretative hypotheses about processes within particular taxa operative at certain spatio–temporal scales. General classifications applicable to all kinds of taxa, caused by similar factors will hardly, if ever, occur. Thus, this aim is not to obtain a natural classification in a taxonomic sense (e.g. Simpson, 1961), but to generate causative hypotheses. Natural classifications do not apply here because of the multifarious ecological adaptations of each individual species, often resulting in parallel or convergent adaptations in both closely and distantly related taxa. The variety of responses to ecological conditions among taxa and of their history, ecology, and time-lags results in an infinite number of classifications. As different classifications result from different taxa, taxonomic levels, and spatio–temporal scales, biological processes of all taxa following individual courses at various spatio–temporal scales, the endless variety of classifications does not imply any criticism of biogeographical classification. It is the hypotheses derived from them that, when tested, are corroborated or falsified. Classifications are not right or wrong, only useful or not. Mayr's (1965a) arguments are consistent with a heuristic use of biogeographical classifications.

Cluster techniques represent one approach for finding biotic discontinuities; if these are not found, the ranges may be randomly distributed. A more analytical way of looking for possible biotic discontinuities is to estimate whether taxa have a random or a clumped distribution. This approach requires knowledge about the spatial structure and dynamics of ranges of species, genera, families, or other taxa, and about spatial trends in the taxon density and properties. This is the subject of Part II.

Before entering that discussion, we look in Chapter 5 at another approach for generating causal hypotheses, ordination.

5

Classification and ordination

Classification techniques, including cluster analysis, are tools for generating biogeographical hypotheses. But they are not the only means of doing this. Ordination or scaling techniques serve the same purpose and can be more useful in this regard than classification. Their greater predictive power arises from representing biological variables directly and hence historical and ecological variables indirectly by means of principal axes or latent variables, although this latter representation is not guaranteed. Moreover, sample scores as ordination results can themselves be ordinated using trend-surface analysis to describe trends in spatial variation of the suggested causal variables (cf. Goodall, 1954; Kooijman and Hengeveld, 1979).

The ideas here are twofold: firstly, spatial variation is continuous and, consequently, cannot be analysed by cluster techniques, based on models that assume discontinuous variation; secondly, several techniques can be used in combination to describe spatial variation. These techniques are known as ordination or scaling techniques as their function is to order geometrically multivariate data, not to partition them into clusters.

This chapter is broadly divided into two. The first part discusses the relative efficiency of the conceptual modes of clustering and ordination techniques. The second part contains an application of some techniques to biogeographical patterns and compares the results with those from classification.

The discussion of ordination is not meant to be exhaustive; it introduces the several lines of reasoning behind the calculations. I shall not discuss factor analysis, reciprocal averaging (i.e. correspondence analysis), or detrended correspondence analysis, nor variations within these techniques. For a recent survey of these techniques, see Ter Braak and Prentice (1988).

64

Efficiency of the models

Classification of biotas to generate hypotheses about the causes of geographical patterns ideally finish with a table such as Figure 14*a*. Here, columns represent the observational locations, and rows the taxa. In this figure locations and taxa have been arranged so that their presences follow the diagonal from the top-left to bottom-right; the off-diagonal corners at the top-right and bottom-left are empty. Furthermore, the distribution of presences is step-wise, resulting in a number of rectangular blocks.

Because of this step-wise distribution of presences this representation shows that these data can be classified; the model of classification is said to be discontinuous. When the cells near blocks 1 and 2 are partially filled, these two can be fused on a higher level of variation, block (1,2), separate from block 3. In this case the model to be used should not only be discontinuous, but also hierarchical.

A representation in which an ordination model applies looks different (Figure 14*b*). Here presences are not distributed step-wise along the diagonal, but form a continuous band, and hence a continuous model applies. By the absence of discontinuities no blocks can be taken together at the various levels of variation and the model is non-hierarchical.

The two underlying models therefore differ in their assumptions concerning the continuity or discontinuity of variation and, consequently, non-hierarchical versus hierarchical arrangement of data. Their function is the same: arranging presences to suggest causal hypotheses about patterns of geographical dispersion of distribution ranges. To judge which model most efficiently generates such hypotheses, we must consider Figures 14*a* and *b* more closely.

By assuming that the distribution along the diagonal is step-wise, two other biogeographical assumptions are implied. First, we assume that several, usually taxonomically unrelated, taxa respond similarly relative to the supposed causative factor, but differently from responses of other taxa. This assumption is unlikely for continuous variation in the causative factor. Second, we assume that all ranges are in equilibrium with the local intensities of the causative factor. Therefore, the taxa are assumed to redistribute and do so that quickly that, on the time-scale of observation, no time-lag can be observed when the geographical pattern of intensities alters. When time-lags occur, and when they differ among taxa, discontinuity decreases, and hence the extent of continuity increases. As climate changes perpetually, ranges continuously redistribute with rates depend-

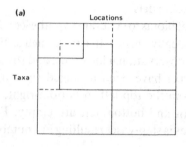

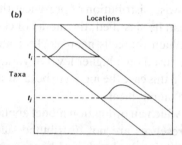

Figure 14. Schematic representation of a data matrix rearranged according to the results of the application of a clustering technique (a) or an ordination technique (b). For taxa t_i and t_j, the optimum-response curves are indicated in (b).

ing on biological properties. Therefore, continuous models are likely to make better sense, unless arguments against continuity exist.

In general, geographical variation will not be discontinuous, and ordination techniques will thus be more useful in generating hypotheses. Constructing classes before inspecting the arrangement of presences along the diagonal can divert attention from possible, overall trends in this arrangement, resulting in a classification being represented as its only end result. Ordinations consider the main effect firstly and discontinuities secondly, if these are found at all.

From this general and abstract methodological viewpoint, ordination techniques are more efficient in generating hypotheses than classification techniques (see Goodall (1986) for statistical arguments). But Figures 14a and b also shed light on recent criticisms of biogeographical classifications such as homogeneity and stability of biogeographical units, sharpness of their delimitation, and hierarchical or non-hierarchical nature of geographical variation of distribution ranges. Much of the debate seems irrelevant from the viewpoint of ordination. Instead of being disputed, they are points of interest, highlighting properties and traits of taxa that respond to changes in environmental conditions.

This chapter discusses stability of ordinations, not only to criticize their results, but also to show the relevance in the analysis of long-term biotic adaptation to temporal instability in environmental conditions. Ordination techniques are not only more efficient and general than classification techniques, they suffer less from technical restrictions in interpretation.

Methodology of ordination

Ordination as a geometrical data representation assumes that a gradient of environmental conditions exists among the sampling sites to which all taxa respond differently. Thus, if the taxa are evenly spaced over this gradient, a continuous band of presences in Figure 14*b* can be seen. If, on the other hand, all species respond similarly, their presences will all cluster somewhere within Figure 14*b*, that is either at some point on the diagonal, or at some distance from it. If they respond differently, but are not equally spaced along the diagonal, this unequal distribution reflects either discontinuities in the gradient or in the responses of the taxa. In neither case do ordination techniques strictly apply; in the first nothing can be done and in the second the data can be ordinated and classified.

Of the various criteria enumerated for classification, only those of similarity and dissimilarity in biotic composition are applied. Often one calculates the covariance or correlation between biotic composition of locations, but these are actually measures of dissimilarity or, more particularly, measures of Euclidean distance.

The procedure applied by ordination techniques is as follows (cf. Pielou, 1984). For two taxa the abundances are plotted as coordinates in the two-dimensional plane they span. Next, one constructs a line or axis representing the direction of the greatest variation and then one perpendicular to it, representing the smallest variation. Finally, the two new axes or principal axes, are shifted and rotated so that they replace the original two-dimensional coordinate system. The point of intersection of the principal axes now becomes the origin of the new coordinate system. The amount of variance of the axes is expressed by a parameter called an eigenvalue. The value of each location where both taxa are represented by particular abundances is called a score. This procedure, visualized for two dimensions, can be generalized to m dimensions for m taxa, or n dimensions for n locations. Graphical representations usually depict the coordinates of points in this multidimensional space in two dimensions only, for which any combination of principal axes can be selected. Often eigenvalues decrease rapidly so that only the first two, three, or four axes are required to account for the greatest part of the variation. The rest is then considered as variation due to random error or noise.

Figure 14*b* also shows that taxa can respond non-linearly to some gradient among the locations: the taxa occur along a certain part of the gradient and not in localities at either side of it. In those cases where local abundances are known instead of presences and absences, it is usually found that abundances are highest in the central part of the trajectory along the gradient and taper off towards both sides. When the samples comprise a complete unimodal response curve, at least two, partly overlapping curves are required to distinguish between similar abundances at the left- and right-hand sides of the curve. If insufficient information is available, the two tails of a unimodal response curve cannot be separated from each other and the two sides of the curve cannot be unfolded (e.g. Coombs, 1964). Apart from ordinating unimodal response curves, one can do the same for polynomial curves, resulting in polynomial ordination (e.g. Phillips, 1978). Unimodal curves like these reflect the numerical response of species to various intensities of a causal factor along the gradient and violate a basic assumption of almost all ordination techniques, namely linear responses. The effect of this violation is to distort the configuration of points in graphical representations of ordination results, known under various terms such as the arch or the horseshoe effect. Different ways of correcting for this distortion have resulted in many ordination techniques.

One part of the methodology concerns descriptive analysis of data, and the other concerns interpretation and testing interpretations. For the latter we will discuss only the first, data interpretation, as the second, its testing, is so determined by the nature of the hypothesis that no general account is possible.

Interpretation of ordination results is done by comparing the properties of points representing locations or taxa at opposite ends of an axis to find the relevant property. When many points share the same property, be it negative at one end of the axis and positive at the other, the axis is suggested to represent a physical factor determining the sequence of the points. This procedure can be repeated for each axis separately. When the sequence of points is distorted because of non-linear relationships, the sequence can be read from the angular distances between points along the resulting horseshoe-like configuration.

For spatial data it is often useful to map the scores for each axis separately according to the coordinates of the sampling locations, thereby facilitating hypothesis-generation and subsequent testing. One can use scores in trend-surface analysis (e.g. Kooijman and Hengeveld, 1979). This analysis is a regression analysis of two dimensions, where the dimensions represent the spatial coordinate system of the observations. Values of the scores are plotted along a third axis perpendicular to this plane. Sometimes it is useful to start from a three-dimensional, spherical surface to allow for altitudinal level or for curvature of the earth's surface (Stehli, 1968). As in one-dimensional regression analysis, the data can be described by linear, quadratic, cubic, or higher-order polynomials. To interpret the pattern, one can look at the trend or shape of the surface representing the data, or at the spatial pattern and intensities of the residuals (Stehli, 1965). In principle, the same idea, though using different techniques, can be applied to species ordinations for comparing the species' geographical or ecological similarity with that of their taxonomy (Gause, 1930).

Interpretation of ordination and trend-surface analysis results are easiest for untransformed data. For transformed data one has to interpret both the ordination and the transformation at the same time. It is even more difficult to interpret, and consequently to test, results of a combination of an ordination technique with trend-surface analysis on transformed data. It is therefore advisable to use statistically sound techniques as principal components analysis or reciprocal averaging (correspondence analysis) in preference to other techniques.

Non-metric multidimensional scaling as an ordination technique does not select particular parts of the original multidimensional space, but warps it. Thus, rather than describing the variation in this space by projecting several axes into it to account for a decreasing amount of variance, it constructs a plane or a higher-dimensional volume such that the configuration of points is affected least. The original space is therefore not kept intact, but replaced by a new plane or hypervolume with the smallest warping. The measure of warping is called 'stress'. This technique can be applied in two-way (ALSCAL), three-way (INDSCAL) or multiway matrices (cf. Kruskal and Wish, 1978) and is particularly useful in the analysis of spatial data (see Holloway and Jardine, 1968, for an application).

Scales of variation and concordant variation

Variation in character state of several characters is discordant when they vary independently in space. Discordant variation is usually, though not exclusively, defined in morphological terms and occurs when (1) trends of variation have a different orientation, and (2) they have a different spatial extension.

Trends with different directions and spatial extensions form the subject matter of ordination, rather than that of classification, as different axes of variation describe such trends, together with their scales of variation. The same phenomenon occurs in coincidence patterns of ranges. Usually, spatial covariation among taxa will be limited, their distribution ranges overlapping only partly. Moreover, some ranges may be extensive, whereas others are restricted. Finally, sensitivity to certain ecological factors can differ among taxa, or they can have different routes of migration. The analysis of discordant spatial variation belongs therefore typically to the realm of ordination, whereas classification is that of concordant variation of taxa.

Concordant and discordant spatial variation are extremes of geographical variation in biotic composition; usually one finds a mixture. Therefore, one often carries out a classification or an ordination, whereas in others one should apply a combination. The more concordant the spatial variation, the better it is to apply a classification technique; the more discordant it is, the more appropriate an ordination technique becomes. But both analyse scales of variation whose effects are superimposed; in classification these effects are expressed by the hierarchical levels (nodes) in a dendrogram and in ordination they are expressed by the various principal axes.

Combinations of classification and ordination

In ordination, one usually interprets results in terms of causative factors, whereas classifications are often considered as ends in themselves. One may wonder if (1) the hierarchical levels (nodes) in a classification can also be causally interpreted, and (2) the interpretation is the same as that of axes defined by ordination. In other words, is it possible to combine classification with ordination to define structures on several scales of variation, so that the importance of the causative factors are weighted for the various scales?

Ter Braak's (1986) combination of TWINSPAN with discriminant analysis is a potentially useful tool here. TWINSPAN defines axes of variation along which the variation is divided for each axis separately. This technique first ordinates the data by reciprocal averaging for the first axis and then divides the points along this axis into two. This procedure is repeated for each of the two sets of points, and so on (Hill, 1979). Simple discriminant functions are constructed that characterize differences at each node. They also classify external attributes according to their frequency of occurrence at each node, allowing the division at each node to be interpreted causally. Because TWINSPAN calculates the axes defining the hierarchies independently, one cannot tell what part of the variance in the data is accounted for by a certain factor. This seems more serious than it is, as the amount of variance accounted for by a certain factor always depends on the scale and intensity of sampling. The advantage of TWINSPAN is that problems of non-linearity of relationships between variables and causative factors are minimal.

Ter Braak (1986) applied his procedure to qualitative data of bird species in 82 Dutch heathlands. First, in two steps he classified by TWINSPAN the heathlands into four, each group being characterized by a particular species combination. To this he applied simple discriminants to characterize these groups in terms of a set of known variables. The results show that the highest frequencies of, for example, the reed bunting are found on heathlands in the northeast of the Netherlands, larger than 100 ha and containing moorland pools. The redshank is almost totally restricted to these heathlands. Other species, however, such as the linnet also occur on large heaths, although they are not restricted to wet heaths and may also occur on large, undulating heaths.

Another combination of ordination and classification procedures first defines a subspace within the original *n*-dimensional space after applying an ordination technique, and then performs cluster analysis within this subspace. This procedure has three advantages: (1) by excluding axes representing noise, clusters can be defined more sharply, (2) because of excluding noise, it confines the analysis to the subspace containing relevant information, and (3) clusters can be calculated for that subspace with a specific, chosen biological interpretation. Next steps can be the calculation of statistical differences using discriminant analysis such as Mahalanobis's D^2 to estimate the significance of the clusters found, or that of discriminant functions, quantifying the importance of the variance-different variables for the differences found. So far, this combination of ordination and classification has not been applied in ecology or biogeography (cf. Hengeveld, 1985c, for discriminant analysis in a reduced factor space).

Stability of ordinations

One criticism of biogeographical classification is that the units obtained are not homogeneous and stable. Lack of homogeneity does not apply in ordinations, because they are specifically applied when there are no discrete units. But one may seriously question if the results of an ordination are stable through time.

Biotic composition of an area changes when distribution patterns change, particularly when ranges of taxa shift independently. Ordinations and their interpretation only hold at the scale of temporal variation of its constituent taxa, which should make the results and their interpretations uncertain and anecdotal. Just as in classification, the central question is thus what we want to achieve by ordinating our data, as the results are not ends in themselves but, as in classification, means to an end. This end is a better understanding of spatial behaviour of taxa in response to environmental variability. When conditions have changed during the observational period, shifts in distribution of taxa confuse these responses and may alter the nature of the trend found.

This could be a drawback of such data sets, but it can also make them more informative for hypothesis generation than when distribution ranges have been stable. Shifts in distribution ranges show more sharply what factors could determine geographical patterns and processes within and between taxa. When species ranges are constant, for example, one could think that their internal structure is static, despite climatic changes. Such species could be physiologically flexible, or unable to invade other communities, because demographic processes of interference would prevail over abiotic ones. This reasoning does not hold when species ranges are not static. They are physiologically not flexible enough, or communities are less closed than expected from stable distribution patterns. Instability of ordinations itself is informative, because it allows us to choose between several interpretations; it is not a drawback that we should try to avoid by choosing another data set. This view contrasts with Kempton's (1981), who recommends using composite species similarity measures or measures for species' relative abundances because of the greater stability of results compared with those from individual species. I think that rather than obtaining a description of the organizational level of communities composed by species of which the names, and hence their biological idiosyncrasies, have to be discarded, it is the biology of individual species that has to be understood.

An application of classification and ordination

Van Balgooy (1971) classified presence–absence data of 1666 phanerogam genera for 36 islands or coherent island groups in the Pacific Ocean into a qualitative, hierarchical system. He also examined a matrix

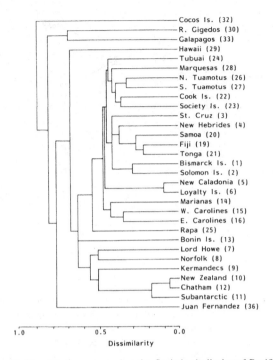

Figure 15. Dendrogram expressing the floristic similarity of Pacific islands
(after Van Balgooy, 1971) (numbers indicate the numbering of the islands
during the analysis).

of Kroeber's (1916) similarities. The result was the recognition of a
trapezoid, equatorial part, bordered by smaller, less important categories
in the north, south, and east. The central area was subdivided into two
subgroups. However, Van Balgooy (1971) did not interpret his classification, although he mentioned two existing theories about the origin of the
Pacific flora. It remains unclear whether several processes explain the
different hierarchical levels and what kind of processes these could be.
Interpretation of a dendrogram obtained from this matrix is also difficult
(Figure 15).

   In a minimum-spanning tree constructed from these data (Figure 16),
the equatorial islands again take a central position and the northern,
eastern, and the Central and South American and the southern islands a
peripheral one. In broad terms the minimum-spanning tree resembles
Van Balgooy's (1971) classification.

   Holloway (1979) applied single-linkage clustering to Van Balgooy's

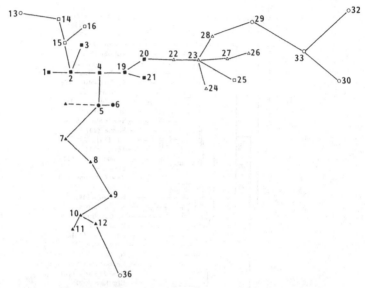

Figure 16. Minimum-spanning tree constructed from the highest floristic similarities of Pacific islands (after data in Van Balgooy, 1971) (numbers refer to those in Figure 15).

(1971) data and recognized five categories, which coincide with Van Balgooy's categories New Zealand, Micronesia, New Caledonia, Western Malaysia and Eastern Malaysia. The remaining islands were recognized as distinct categories. Holloway's (1979) results obtained quantitatively thus reflect those obtained intuitively by Van Balgooy (1971). As most island floras are, in his view, a disharmonic blend of elements derived from many source areas, Holloway did not consider his classification very informative. However, depending on our aims when classifying, this may be just the information we want to draw from classifications!

Interpretation of these results is possible, however (Figures 17 and 18). East and West Malaysian islands may contain equatorial floras and the floras of Micronesia and New Caledonia and the Loyalty Islands lying at some distance from the equator may be more subtropical. Still farther away, the group of islands Van Balgooy classified as New Zealand are more related to the Bonin Islands and Hawaii in the Northern hemisphere than to the equatorial islands closer by. They are also related to the equatorial islands off the American coast, which are surrounded by cold upwelling water and, consequently, may be relatively cool.

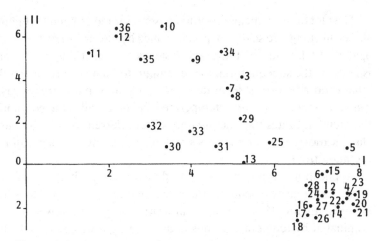

Figure 17. Floristic composition as ordinated with principal coordinates analysis for the first two axes (after data in Van Balgooy, 1971) (numbers refer to those in Figure 15).

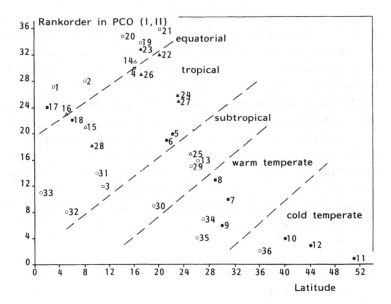

Figure 18. Rank order of the ordinated floristic compositions of 36 Pacific islands plotted against the latitude of the islands (after data in Van Balgooy, 1971) (numbers refer to those in Figure 15).

The plot of the principal coordinate scores for axes I and II (Figure 17) shows an elongated scatter of points, which by plotting their rank number against the latitude of the islands, shows a trend from the equatorial islands via the subequatorial to the temperate islands. Hawaii and Bonin Islands do not form a separate, northern group, but, surprisingly, are found among the southern temperate islands. Finally, the continuity of this trend indicates that the discontinuities inferred from the qualitative classification and cluster analysis may be non-existent or subordinate to the main trend.

Application of non-metric multidimensional scaling (Kruskal and Wish, 1978) gives essentially the same picture as principal coordinates analysis. Figure 18 shows an oval scatter of points, towards the left dominated by temperate islands and to the right by equatorial islands. The points at the left and above the horizontal axis represent eastern islands, and those below this axis represent western islands. Moreover, no clear distinction can be made between northern and southern islands, either in those along the equator, or in the eastern and western temperate ones. Thus, the floral composition of Hawaii is more related to that of Easter Island, or the Desventuradas Islands in the Southern Hemisphere than to the Bonin Islands in the Northern Hemisphere. But the Bonin Islands are rather similar in their floral composition to Rapa Island or to New Caledonia in the Southern Hemisphere. Again, the most southern island in the east is similar to the most southern islands in the west.

Therefore, two tendencies seem to dominate the similarities of the floral composition among the islands: (1) an equatorial-to-temperate trend, and (2) an east-to-west trend. The equatorial-to-temperate trend might be explained by ecological factors other than distance, and the east-to-west trend by distance acting as, or involving a barrier. This barrier is weakest at both ends, namely in the temperate regions and the equatorial regions. These tendencies make classifying these floras difficult. This is remarkable, as it would imply that 78% of the variation represented by the ordination is not accounted for by classification. Concentrating on discontinuities (classification), rather than on overall trends (ordination) in his data may have prevented Van Balgooy (1971) from formulating explanatory hypotheses.

Conclusions
Data on the biotic composition of various geographical sampling locations can both be classified and ordinated, according to the information one is most interested in. When classifying, one is interested in

discrete groups of locations or taxa, whereas in ordination one emphasizes trends among these locations or taxa. However, usually data sets neither consist of assemblages of discrete clusters of biota that do not integrate, nor are they perfectly continuous. Consequently, both classification and ordination techniques describe only certain aspects of the data, ignoring other aspects. But they agree in that they generate hypotheses about the causes of the structure in the data. The choice of technique is partly subjective, reflecting the student's interest, and partly objective, depending on the degree of efficiency of the data representation.

Summary of Part I

Part I has been concerned with the first stage of analytical biogeographical research, namely the methodology and techniques to aid in hypothesis generation. The ways of doing this are varied. One can look at differences between two or more areas distinguished by adopting *a priori* criteria, or one can first classify or ordinate the areas on the basis of their biotic content. The first approach adopts a variety of criteria, the best known of which is that of the distinction between continental plates. Because the latter includes a historical argument, it carries a risk of circular argument, a risk that can also be present in other *a priori* criteria.

Classification and ordination techniques are designed to avoid this problem. Ideally, one counts the number of taxa, or even the number of individuals per species, in the squares of a grid system or sections of a linear transect. Then the counts for all these squares or sections are compared with each other. Classification adopts a discontinuous model, whereas ordination techniques are based on continuous models. The first kind of technique gives information on concordant spatial variation of the taxa concerned, and the second kind on spatial trends resulting from discordant variation.

The hypotheses generated include two different types, those formulated in terms of the past history of the areas or taxa, and those formulated in terms of ecological variables. The first type assumes that the ecological conditions have remained more or less the same through time, or that the change has not caused the species to die out. In the latter case the individuals may prove to be flexible or adaptable enough to attune to the new conditions or, when they are not, that the range moves towards regions with more optimal conditions, thereby often lagging behind the changes to some extent. Then the lag can also be categorized as a historical factor, although the new lag pattern may not show any overlap with that at the start of the spatial shift. Obviously, the two types will hardly, if at all, be found in a pure form. The sharpness of the distinction between historical and ecological explanations will depend on the degree of flexibility or spatial adaptability and these in their turn on the spatio-temporal scale chosen. Results may differ considerably at another scale.

All the methods discussed are heuristic, that is their results are preliminary only. They contain the minimum number of assumptions, and except for similarity coefficients and cluster techniques, they are relatively robust, implying (1) that different techniques applied to the same data give more or less the same results, and (2) that deviations from the assumptions made do not distort the results too much. Finally, they are not predictive in the sense that they can predict results from an analytically derived or constructed model. The pattern they may expose in the data generates hypotheses or predictions that have to be tested only in the next phase of research. But at this point we have to be cautious, as the pattern can be artificial, resulting from a particular way of sampling, a certain choice of technique, a certain assumption as to the underlying sampling distribution, certain relationships of the taxa mutually or with causative variables, or a certain dynamic behaviour and size of time-lag, etc.

As hypothesis-generating techniques, their application is limited, because they assume that no hypothesis is yet available. In those cases where one or more hypotheses are available, these can be tested straight-away. It is wise to apply one or more of these techniques even in those cases, to make sure that the supposed causative process is the most important one, and also that it is operative on the spatio-temporal scale investigated. When applying these techniques one can with care avoid circular reasoning and artificially imposing *a priori* ideas on the data.

Discovering a pattern in the data this way requires a certain knowledge of the idiosyncrasies of taxa and regions, for example which are good dispersers, which prefer a certain temperature range, or which are found in remote areas or of a certain minimum size. The hypothesis generated involves a generalization of this knowledge from a limited number of taxa or regions to all that show some relationship to this pattern. This relationship is shown by their relatively short distances in a graphical or geometrical representation of the numerical results. But 'relatively short distances' implies that it should be possible for relatively great distances also to occur. This means that one should not select homogeneous regions before starting such procedures, or choose more or less similar taxa, that, on top of this, are perfectly static over the period investigated. On the one hand, the procedure to make the data set homogeneous in these respects would introduce *a priori* information into the analysis, again involving the risk of circularity of reasoning. On the other hand, it would take away the possibilities of recognizing the patterns anyway. Both preclude, to a smaller or larger extent, further analysis.

Stating the idiosyncrasies of several taxa or regions should be known also implies that the procedures should be used to allow the use of this knowledge. Synthetic parameters such as diversity indices are therefore of little use. Instead, it is better to perform two kinds of comparisons on the analysis, that of geographical regions relative to each other, as well as that of taxa relative to each other. The most informative procedures in this connection are those that integrate results of both kinds in a biplot analysis or a joint plot as in correspondence analysis.

The philosophy basic to all methodology of the analysis of patterns of coincident distributions is that the results are not ends in themselves, but rather means to an end. Therefore, we should avoid any technique or methodology that does not lead to the generation of testable hypotheses. If part of the information cannot be tested because of lack or shortage of information, as often happens in effects from processes that occur in the past, this part should be separated in an early stage of research and as sharply as possible from that part that does generate testable hypotheses. Any further progress in all sciences, including biogeography, depends on progress by testing more and more newly generated hypotheses.

In this vein, we will first approach the analysis of the geography of biological traits irrespective of taxonomic identity in Part II, and then the geography of patterns of distribution of identified taxa in Part III.

II

Geographical trends in species richness and biological traits

Part II considers geographical trends in species number over large, even global areas, or in more restricted regions (Chapter 6) as well as trends in biological properties (Chapters 7 and 8). Chapter 6 gradually narrows the spatial scale from large, global trends in broadly defined taxa to trends within genera in restricted areas. The sections within Chapters 7 and 8 cannot be arranged this way; however, the distinction between these chapters reflects, in part, differences in spatial scale. Chapter 7 covers global or continent-wide trends or patterns of variation, irrespective of taxonomic identity, whereas Chapter 8 considers finer-scale patterns, often within species.

The topics in each section are varied, as are the taxonomic level and identity. Putting them together will therefore be surprising, unless the phenomena are considered as geographical responses to broad- or fine-scale environmental variation. Here again, the pattern, as such, is not of primary interest; what is of interest is the patterns as a spatial expression of species' ecological responses to environmental variables.

One can wonder if the taxonomic level is too high and the spatial scale too broad to warrant our attention. Here again, the interest depends on the scale of variation of the phenomenon concerned. Also, depending on the area, all individuals of the species in the taxon concerned, whatever its level, are subjected to the same environmental variables. Studying one species only, even if we consider its total range, does not allow us to discover patterns and responses such as gradients in leaf form. The discovery of all environmental variables to which species respond requires the study of phenomena at various levels of integration and on all possible spatio-temporal scales.

In the choice of levels of integration several criteria are used to classify taxa into categories. To this end the relative representation of these categories is estimated over the area investigated. We can then try to find environmental variables whose intensity covaries with the categories' relative numbers. In doing this, we are postulating hypotheses about the way species within the category respond relative to some environmental variable(s). Finally, we should try to find a means of testing this hypothesis by understanding the mechanism underlying it.

In many cases the stage of hypothesis-testing has not been reached. Instead, there is often still uncertainty as to the reality of the phenomenon or its possible causation. Most, if not all, of these trends have been described qualitatively rather than tested statistically, partly because many of them have long been known, sometimes from the turn of this century or earlier. Another reason is that statistical techniques for testing spatial trends or patterns are still poorly developed. The following three chapters will thus not discuss the statistical reliability of the patterns, but rather take their existence for granted. Wherever possible, I discuss literature that makes the patterns comprehensible or that contains results of experimental tests.

6

Geographical trends in species richness

Distribution ranges can be understood in terms of historical factors or ecological adaptations. The latter reflect the importance of species withstanding environmental variability; sooner or later they will die out if they cannot tolerate such variability. This chapter maintains this ecological emphasis, although it concerns a different biogeographical phenomenon, the geographical distribution of numbers of taxa per unit area, or taxon density. I will relate this density to ecological conditions, present and past, or to the systematic affinity of the taxa.

Again we face the problem of how to select the appropriate spatial scale and taxonomic level. Some trends concern global clines in species number, irrespective of their identity; occasionally the number of species will even be added together! In contrast, other trends concern, for example, gradients in ecological relationships between closely related species, occupying part of Europe and interpretation varies accordingly. Obviously, too many and ecologically too divergent species make it hard to choose between several plausible hypotheses to explain observed trends in taxon density, or to test the applicability of particular explanations. It is difficult enough to estimate how far densities can be explained when only a few closely related species are studied. However, this difficulty is not a drawback connected with taxon density of closely related species, but a problem basic to our understanding of biogeography.

Latitudinal trends

The tropics are richer in species than biotas at higher latitudes, a phenomenon that Fischer (1960) described quantitatively and for which he presented several explanations. However, he did not describe richness gradients in detail; much more information is needed to provide detailed descriptions necessary for testing causal hypotheses. Many later papers only suggest possible causes of the gradients described, thereby adding little information. It is, moreover, often necessary to restrict study to variation on a finer spatial scale, such as one continent or a part of it, or to numbers of species of a taxon of a relatively low rank. On the other hand, in doing this, one can easily fall prey to reductionism, losing sight of the phenomena concerned.

Stehli and co-workers studied global past and present distributions of species of taxa of a very high rank, such as Foraminifera and Mollusca. In these studies they counted, at many sampling sites, the number of species. They then applied trend-surface analysis, largely to estimate the geological pole location. First, Stehli (1965) and Stehli and Helsley (1963) mapped the numbers of planktonic foraminifer species and showed that species number is highest in oceans at low, equatorial latitudes and lowest in temperate and arctic regions. They suggested that temperature is the factor explaining this global distribution of species richness. Yet, regional species number may deviate considerably from that described by the trend-surfaces, being either too high or too low. The mapped residuals relative to a particular surface (Stehli, 1965), show that positive deviations coincide with warm waters and negative ones with cold waters. This suggests that ocean current patterns are a secondary factor explaining the geographical distribution of species numbers, superimposing that following a latitudinal temperature trend.

This conclusion differs or applies only partially to other taxa. For example, numbers of benthic foraminiferal genera sampled at 104 locations (Durazzi and Stehli, 1972) are highest near the equator and lowest at higher latitudes. To explain this, Durazzi and Stehli estimated for each location the average generic age. Disregarding location longitude, these generic ages were plotted against latitude, which results in a hollow distribution and indicates that generic age is lowest near the equator and highest at the Poles (Figure 19). With longitude added to the locations the trend-surface of the generic ages shows the same latitudinal trend. Since the calculated North Pole (87°40′N, 38°12′W) almost coincides with that calculated from sea-surface temperatures (89°N, 120°E), Durazzi and Stehli suggest that, as in bivalves (Hecht and Agon, 1972), speciation is affected by temperature. Speciation may be highest in warm tropical waters and decrease towards higher latitudes. The young, tropical species may not have had time to disperse to these latitudes.

Thus, two explanatory hypotheses can be formulated, an ecological hypothesis and an evolutionary one, which itself may be ecologically dependent. This is suggested by possible relationships between speci-

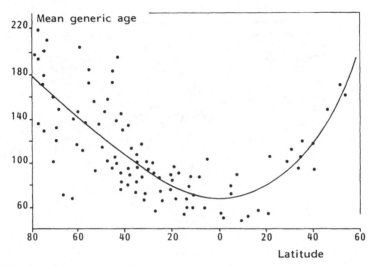

Figure 19. Mean generic age of benthic foraminiferal genera plotted against latitude (after Durazzi and Stehli, 1972).

ation and temperature, and between generic age and distance from the equator to the Poles.

Longitudinal trends

Stehli (1968) also described gradients connecting longitudinal nuclei of high species richness in several taxa, here called 'richness province'. Sometimes only two richness provinces are found in terrestrial taxa, those of South America and South-East Asia, Africa having relatively low species numbers. This occurs, for example, in turtle species (Pleurodira), in parrot genera (Psittaciformes), in mosquito genera (Culicidae), and palm genera (Palmae). In marine taxa this occurs in the bivalve family Pinnidae.

Stehli (1968) suggests that marine richness provinces (Figure 20*a*) are delimited by temperature barriers; he does not explain terrestrial prov-inces, or the frequently observed absence of an African province. Richness provinces in terrestrial taxa may be delimited by water barriers (oceans), temperature, or humidity, the effectiveness of which will be less for taxa with mobile species than for immobile ones. Stehli (1968) emphasizes that the relative weakness of the richness-province concept is not solely a function of present environmental conditions, but it may also reflect historical events, such as climatic events and the area and period of

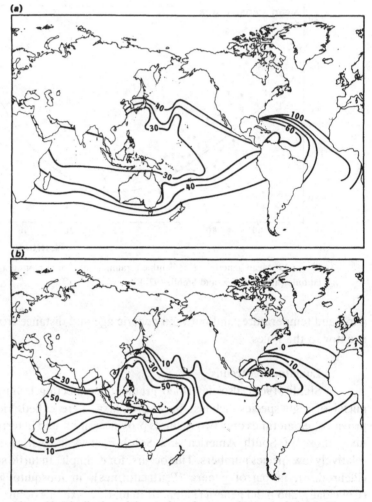

Figure 20. Distribution of the number of genera of hermatypic corals (*b*) and their average generic age in millions of years (*a*) (after Stehli and Wells, 1971).

origin of the taxon. The unexplained relatively low richness of African taxa may perhaps reflect the more continental climatic conditions when Africa was situated centrally within Pangaea. Another explanation may be its physiography with a high, craton-like peneplain.

Generic age may also differ in other taxa, both within and between oceans. Stehli and Wells (1971) estimated the generic richness and age of hermatypic corals from 63 locations. Three richness provinces are dis-

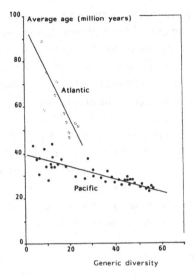

Figure 21. Average generic age of hermatypic corals plotted against generic diversity for the Atlantic and Pacific oceans separately (after Stehli and Wells, 1971).

tinguished, in the Indian, West Pacific, and West Atlantic oceans (Figure 20*b*). High richness nuclei in western parts of the oceans are also found in other taxa, possibly reflecting relatively high sea-surface temperatures. On average, genera within or near a nucleus are younger than away from it. Also, Atlantic genera are, on average, older than those in the Indian Ocean and West Pacific (Figure 21). Moreover, the number of genera in the Atlantic lacking a geological record is lower than in the other oceans. The highest number of genera in the Western Atlantic is much lower than in the other oceans, which Stehli and Wells (1971) assigned to its smaller area: the smaller the area the lower the generic richness. Mean annual sea-surface temperature does not account for the differences, as it is similar for the three areas.

The rate of evolution may thus be highest in the richness centres, and richness itself is possibly related to area. Speciation may mainly take place in these centres, and new species spread in all directions, gradually adapting to less favourable conditions.

Continental trends in avian diversity

In general, one can sample large areas at the expense of spatio-temporal sampling intensity and numbers of species sampled, or, conversely, sample a large taxon frequently and at many localities at the expense of the size of the area. The problem here is one of scale, and all results depend on the choice we make. For latitudinal richness gradients, one can ask, for example, if speciation rate gets higher towards the equator, as suggested for benthic foraminifers and bivalve molluscs, or if tropical families are an ecologically different sample from those at higher latitudes, and their numbers originated by differences in habitat selection. For some reason, this difference in ecological preference may relate to taxon age. For example, Slud (1976) found that the proportion of phylogenetically older non-passerine birds increases northward over North America. This implies that we need to know the biological identity and properties of the taxon to understand richness gradients, rather than relying on results from species counts only that disregard biological information.

North America appears to be large enough to show richness gradients in mammals and birds. But in the case of birds at least, it seems that to study species of all families jointly at this spatial scale is not appropriate to obtain useful, interpretable results. Rather, distribution of species richness should be studied separately for each family, or even at lower taxonomic levels (see, however, Currie and Paquin, 1987).

From a total of 1257 bird species, Cook (1967) counted species number per 95×95 km^2 in a grid laid over North America as an estimate of species richness. A north–south richness gradient was shown to exist, running from low numbers around Hudson Bay (less than 30) to high numbers (more than 480) in the south (Figure 22a). A second gradient runs east–west with high numbers in the Rocky Mountains and low numbers in the east. Species richness also decreases towards the tips of the four peninsulas of south-west Alaska, California, Florida, and Yucatan. Simpson (1964) found the same three gradients in North American mammals (cf. also Connor and McCoy, 1979).

There are several explanations for these gradients (Cook, 1967). Apart from factors possibly operating on a hemispherical scale that determine latitudinal trends and boundaries, local topographical factors may also be important, thereby accounting for the higher species density in the Rocky Mountains compared with flatter areas of eastern North America. This may be due to a greater number of habitats per quadrant in the mountains than in the flat or undulating countryside of the north and east. Apart from ecological effects resulting from topography, east–west differences in species density can also be explained historically. During the glacials north–south shifts of North American biota were complicated

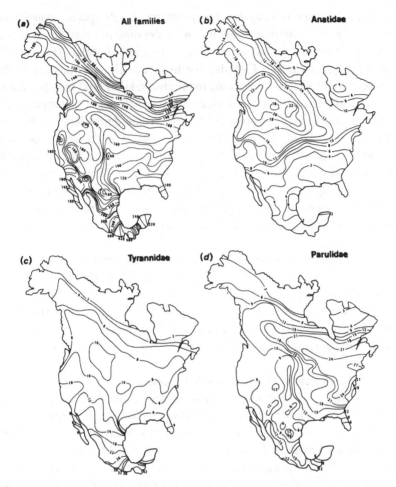

Figure 22. Species richness of North American bird faunas for all families together, and for three families separately (after Cook, 1967).

by temporary ecological isolation of the south-east from the west. The relatively humid south-east is presently separated from parts of the west by a very dry belt, thereby limiting species exchange. Thus, a historical sequence of temporary isolations may also explain the decline in species density towards the east. Finally, the peninsula effect may be due to the spatial dynamics of individual species being restricted by the smaller available space, thereby selecting species that require small areas only (see, however, Means and Simberloff, 1987).

Cook (1967) does not discuss the way ecological factors could influence species density for the gradients of all species considered jointly. He gives maps and slightly more detailed information for some families. Thus, the restriction of ducks and geese (Anatidae) (Figure 22*b*) to central and western Canada, northwestern United States, and the Rocky Mountains may reflect the many rivers, lakes, and marshes there. In contrast, flycatchers (Tyrannidae) (Figure 22*c*) are confined to the warmer, southern parts, possibly because of their dependence on insects. Titmice (Parulidae) (Figure 22*d*) are restricted to the coniferous and deciduous forests of eastern Canada and the United States, possibly because of adaptation to temperate conditions. Finally, the Rocky Mountain topography seems not to affect birds of prey (Accipitridae), possibly because of their predatory behaviour, generally greater size, more extended hunting areas, and relatively varied ecological requirements (Cook, 1967).

Thus, the extent to which general conclusions can be drawn depends on the availability of detailed knowledge of both the history of environmental conditions and their present-day variation, as well as of ecological variation of taxa – in this case bird families – within the whole group investigated. Without such detailed knowledge it is hazardous to make any general causal inferences.

Continental trends in Holarctic plants

So far I have discussed studies based on species counts in several sampling locations or grid squares. These did not give information about parameters of individual species ranges, such as location, size, and shape. In so far as dynamics of species ranges are adopted to explain observed gradients, it is never explicitly central to their explanation. Stehli studied all planktonic and benthic foraminifers jointly, that is hetereogeneous taxa. Species richness of mammals and birds was studied on supposedly homogeneous taxa, but the birds, at least, are heterogeneous at the family level.

Hultén's (1937) study of distribution patterns of Holarctic phanero-gam species is markedly different. He determined the location, size, and shape of the ranges of about 2000 species in Siberia, Europe, and North America. Comparing these geographical parameters, he classified them into several groups, each comprising species with similar ranges. The group range was called an equiformal progressive area (Figure 23) and was explained in dynamic historical terms. During the last glaciation, species may have occurred in a relatively small area of eastern Siberia from which they spread at varying rates. The zone within the equiformal

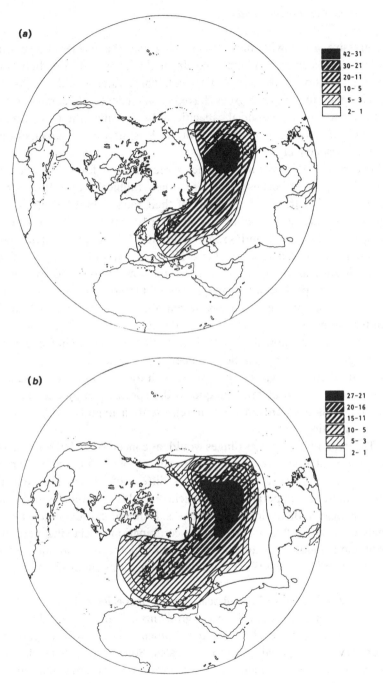

Figure 23. Progressive equiformal areas expressing the number of plant species with a similar presumed area of origin after the last glacial (after Hultén, 1937).

progressive areas with most species suggests the glacial refuge; the decrease in species numbers away from this zone indicates between-species variation in their rate of spread. These rates would vary as the species may, to various degrees, have become genetically depauperated with regard to their dispersal capacities. Hultén called the two extreme rates plastic and rigid for the genetically least and most depauperated species, respectively.

Hultén's (1937) picture of species recolonizing formerly occupied areas is important because of its broad geographical scope. It does not confine itself to the European part of their ranges, but, because of its broadness, puts disjunct arctic and alpine species populations into a new perspective. Some difficulties remain, despite Hultén's long and intimate experience with these species. First, Hultén did not make his classification criteria explicit, leaving us uncertain how he assigned species to particular groups. Instead of mapping individual species, he only showed the equiformal progressive areas, which often overlap greatly. Second, Hultén did not supply information on dispersal capacity, nor on genetic depauperation to show that spatially restricted species are in fact genetically rigid. Concerning genetic depauperation one needs to know species' plasticity before the glacials compared with today. Third, it is assumed that species spread at rates unique to the species and independent of climate. This is interesting, but it represents a hypothesis that badly requires testing.

The dynamics of species ranges would, as conceived by Hultén (1937), be a historical process, occurring independently of climate, rather than an ecological one. It is explained by species' differences in dispersal capacity and genetic depauperation. As such, it adds dynamic aspects to geographical gradients in species richness. It should, however, now be re-analysed statistically, starting with a combination of classification and ordination, followed by testing of the null-hypothesis of no equiformal areas. Then Hultén's (1937) causative explanations should be tested.

Explanations of broad-scale trends in species richness

The solution of problems of gradients in many unrelated taxa, or of a high taxonomic level, is extremely difficult, particularly for patterns over very broad spatial or temporal scales. Some gradients reach even global or geological dimensions. Both taxonomic heterogeneity and a broad spatio-temporal scale make causal analysis of these gradients virtually impossible.

Yet, many explanations exist (e.g. Dobzhanski, 1950; Pianka, 1966), their sheer number probably indicating that their application or relative importance has not been tested. Moreover, though widely different, all the explanations seem plausible at first sight. One explanation (1) is that climate becomes progressively more stable towards the equator, allowing species to specialize or adjust to each other more than is possible at higher latitudes. This allows ecological niches to become narrower, creating room for more species in communities. Moreover, (2) tropical conditions would have been relatively little interrupted, resulting in a long, undisturbed evolutionary process. (3) Rate of speciation would be higher at higher temperatures. Also, (4) biological productivity in the tropics is higher than at higher latitudes because of higher, tropical temperatures. (5) Under temperate and arctic conditions populations would be regulated by harsh, abiotic factors, making their environments relatively stressful. Abiotic factors would be replaced by biotic processes such as competition or predation towards the tropics. Alternatively, a positive relationship exists between diversity and energy input (Currie and Paquin, 1987; Turner, Lennon and Lawrenson, 1988). This can result in a gradual latitudinal shift in species composition related to, say, temperature, which is, moreover, specific for each taxon or ecological grouping of species (e.g. Heggberget 1987). (6) Population densities in the tropics could be kept low by the great variety of predators, diseases, and parasites occurring there, enabling more species with low densities to share a common resource than when their densities are higher. Also, (7) ecological communities in the tropics would be spatially more varied and complex than those at higher latitudes, thus facilitating co-existence of relatively large numbers of tropical species, whereas this would not be possible in spatially more homogeneous and less complex temperate and arctic communities. Finally, (8) the size of the entire tropical surface-area would have a positive effect on total species number, such that local species richness would be higher.

   All these factors are plausible to some extent and some evidence exists to support many of them. Moreover, as expected, evidence exists indicating that these factors and processes do not operate in isolation, but that certain combinations may permit local levels of species richness. For example, Vermeij (1978) showed that, together with a latitudinal increase in richness of marine species towards the tropics, predation pressure also increases. Taylor and Taylor (1977) found, accordingly, that the number of predatory gastropods south of the boundary between continuous productivity and seasonal (summer) productivity at 40°N greatly exceeds that north of it. This implies that a continuous supply throughout the year permits large numbers of predators with specialized diets to exist. Moreover, relatively high predation pressure in tropical parts of the oceans can result in relatively quick mutual adaptations of predator and prey species (e.g. Vermeij, 1982). This agrees with the relationship found by Stehli and Wells (1971) between generic richness and age of hermatypic corals. Regression of age on species richness differs between the Atlantic and Pacific, the latter having the greatest area and greatest richness of coral genera, which are also the youngest. In this case area and annual sea-surface temperature may also be important, correlating negatively with generic age and positively with generic richness.

Explanations in which predation pressure is inferred from morphological parameters (see, however, Abele *et al.*, 1981) and age is directly estimated from fossil evidence, apply only to marine environments. Other relationships such as between area and sea-surface temperature and speciation rate are more difficult to estimate, although they may also play a part. But often, particularly with terrestrial taxa, we have only little information to test the applicability of causal hypotheses; moreover, when information exists, matters usually appear too complex for analysis.

This is also the case with gradients found on a spatially finer scale for taxa of relatively low taxonomic level. We are then often uncertain about possible effects of climatic changes on migration, adaptive radiation, and extinction. We need several criteria to determine possible centres of origin of taxa, if such centres exist. We know too little about speciation processes themselves, and about eco-physiological affinities between species to explain uni- or multicentre species nests. To illustrate the difficulties of interpreting broad-scale gradients, I will briefly look at three examples of species nests, all with different explanations.

Geographic nesting of species

Some restricted regions may have relatively high species numbers, that decrease away from the region (Figure 25). This phenomenon, when it occurs within one taxon (e.g. family or genus) is called species nesting and has been described particularly for plant species. Many other taxa in widely different families share this pattern (e.g. Briggs, 1984).

Species nesting is often thought of evolutionary interest, especially the kind of speciation that may explain it. One possibility equates the area of highest species density, the taxon's diversity centre, with the centre of phylogenetic origin of the genus or family (cf. Cain, 1944). According to Pielou (1979*b*) this is wrong, as it implies that speciation could be

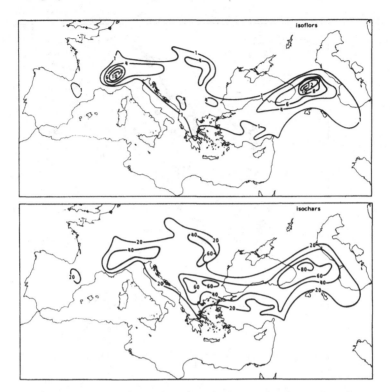

Figure 24. Isopleths of the number of species irrespective the number of characters (isoflors) and, conversely, the number of characters irrespective of species number (isochars) for the species of the subgenus *Calycanthum* within the genus *Alchemilla* (after Rothmahler, 1955).

sympatric, which for theoretical reasons she – with many others – feels is impossible (see, however, Barton, Jones and Mallet, 1988; White, 1978). I feel that the speciation process she favours to explain species nesting, quasi-sympatric speciation, differs only in the scale of observation. One can always adjust or refine this scale so that sympatric speciation either can or cannot happen, thus reducing the study of a biological process to a quibble about definitions. I prefer not to discuss speciation phenomena using these terms, but to describe the processes within the framework of observational scales.

Willis (1922) equated the centre of diversity of a taxon with its supposed centre of origin. According to him, species originate in the centre of diversity, and spread in all directions with time. Species, or taxa in general, with the largest ranges would be oldest, those with the smallest

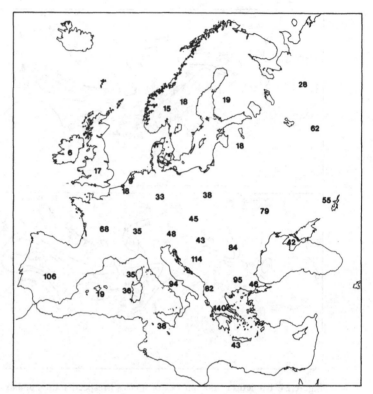

Figure 25. Regional number of silenoid species across Europe (after Thompson, 1973).

ranges youngest. This idea, though heavily criticized (e.g. Fernald, 1926), can also be found in explanations of latitudinal gradients, as well as in Brown's (1957) centrifugal speciation, or the epidemiological and population genetic literature.

The main criticism is that species nesting is explained entirely in terms of the time elapsed since species origination, thus ignoring their ecological requirements and processes, as well as differences in dispersal capacity (e.g. Fernald, 1926). Alternatively, by ignoring effects of time, overlapping of distribution ranges can also be explained by the degree of the species' systematic, and hence ecological affinity; the greater the affinity, the more their ecologies, and hence their ranges, overlap.

In practice these two opposing ideas may be complementary. Apart from systematic and ecological affinity, one should also take the nature

and steepness of environmental gradients, and the environmental variability, into account. Therefore, it can often be misleading to equate
species nests with the area of origin. In his criticism of the area of origin
concept, Cain (1944) enumerated 13 criteria, none of them being water-
tight when considered separately. Additional criteria, such as the inclu-
sion of regions with the highest number of parasites or morphological
diversity, should help in locating this area and in explaining species nests
in proximate terms.

Using morphological diversity as an additional criterion, Rothmahler
(1955) mapped the number of characters per species and then compared
the pattern of locations with the same number of characters, isochars,
with that of the same number of species, isoflors (Figure 24). For species
of the subsection *Calycanthum* within *Alchemilla* the patterns are not
congruent, the difference indicating the discrepancy between centres of
diversity and origin. Rothmahler (1955) considers the Caucasus with 12
species and 80 characters as the primary centre of development, and the
Alps with 10 species and 40 characters as a secondary centre. He did not,
however, interpret the 4 species with 60 characters in the Balkans in the
same way. Rather, he called this a relict area with heterogeneous
elements, implying that adding this morphological criterion is insufficient
to solve the problem.

This example illustrates the difficulty of this approach. This difficulty is
even larger in the *Galium* complex of the eastern Mediterranean
(Ehrendorfer, 1951, 1958). It reflects our ignorance of the spatial
dynamics of species and their adaptive radiation and extinction in the
past, due to glaciations and other, lesser climatic fluctuations. The case of
Calycanthum also shows that more than one centre of diversity may
occur, even within one subsection of a genus (Figure 24). Apparently,
species are not spatially homogeneous and genetically coherent, but are
divided into several, more or less, independent demes occurring in a
mosaic of habitats (Allard, Miller and Kahler, 1978). Species are dynamic
entities with geographically discordantly varying properties. Spatially
discordant adaptation of several ecologically relevant properties that
some species share allows nests to shift in space as systems. It also allows a
continuous spatial reshuffling of species within a species nest relative to
each other without trends in richness being disturbed. The assumption of
dynamic and spatially discordant variation differs from that related to
questions of speciation being, in principle, sympatric, quasi-patric, or
allopatric. Species nests, in their turn, can thus also be interpreted in
terms of degree of concordance of several ecologically relevant species

properties and not in terms of a central area of origin, continually producing new species.

The present section contains three examples of species nests, illustrating the impacts of ecological, historical, and genetical factors, respectively.

The impact of ecological factors on European Silenoidae

The species nest within the Silenoidae (Caryophyllaceae) concerns 23 species ranges (Figure 25). To explain the European part of these ranges, Thompson (1970, 1973, 1975) related their location and size to temperature response during germination, which differs markedly between species. For example, a more southern species such as *Petrorhagia prolifera* germinates at higher temperatures than a northern species, such as *Silene tartarica* and has a wider amplitude (Figure 26). Seeds of *P. prolifera* from France and Hungary vary little in temperature response, in comparison to widely dispersed species, such as *Silene vulgaris* or *S. dioica*. *S. vulgaris* varies greatly morphologically and occurs in a relatively large number of habitats.

Thompson distinguished three types of European distributional ranges: (1) Mediterranean species, (2) species from the deciduous-forest zone, and (3) Russian steppe species. The Mediterranean species germinate as winter annuals at relatively low temperatures (4.5–24.2°C), thus avoiding the dry summer conditions. A period of seed dormancy after ripening prevents germination. Species in the deciduous-tree zone germinate at more average temperatures (11.8–30.0°C). Two groups occur, those germinating immediately after ripening in the late summer, and those in the next spring after a period of dormancy. The steppe species are least restrictive; their seeds germinate over a wide temperature range (6.0–30.0°C), depending on precipitation. They can germinate in the summer immediately after ripening, and young plants overwinter under snow.

Therefore, species with a similar European distribution have similar germinating conditions and may or may not require dormancy before germination. Moreover, germination conditions of local populations of species with extensive ranges, such as *Silene vulgaris*, differ markedly from each other, resembling more restricted, local species. Thus, germination physiology of Silenoidae species seems to be important in determining location and size of their geographical ranges. This overlap of species ranges, due to variation in germination physiology relative to a certain steepness in temperature gradient, explains the pattern in species

(a)

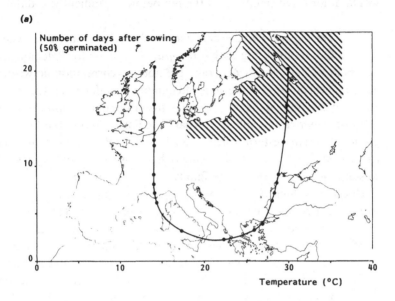

(b)

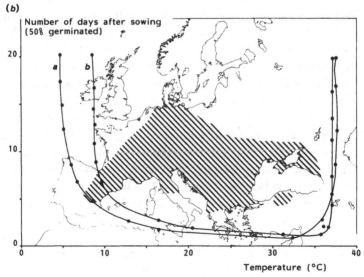

Figure 26. Geographical range in Europe and temperature range of germination for *Silene tartarica* (a) and *Petrorhagia prolifera* (b) (after Thompson, 1970). The two temperature response curves in (b) refer to seeds originating from France (line a) and from Hungary (line b).

richness found. It is likely therefore that the species-richness gradient would differ if the steepness of the temperature gradient was different.

The impact of historical factors on North American Polemoniaceae

Grant (1959) found that the greatest number of Polemoniaceae occurs in western North America. He noted a correlation between the geographical position of the area and number of endemics, rather than its size, and determined the proportion of five tribes of this family in various parts of America to explain this pattern. For example, the Gilia tribe predominates in western North America and the Polemonium tribe more to the east and north, which, taking richness as a criterion, suggests that western North America is the family's centre of origin.

But morphological and cytological properties of all species, together with the environmental history of America give a different picture. Of the five tribes, Cobea, Cantua and, to a lesser degree, Bonplandia, contain most of the primitive properties, and Gilia and Polemonium the least. At present Cobea and Cantua occur in montane tropical rain-forests and tropical woodlands, respectively, which during the Eocene and Early Oligocene had a much greater geographical range in North and South America. Grant assumed that their ancestor occurred in comparable tropical environments. During the Late Oligocene and Early Miocene, when the climate became more arid, this tropical flora developed xeric vegetation in North America and in South America.

In the tribe Gilia, with genera such as *Gilia*, *Eriastrum*, and *Langloisia*, more species developed as desert plants because of a continuing decrease in rainfall during the Late Pliocene, and even desertification in the rain shadow east of the mountain range that, at that time, formed along the Pacific coast. Of these genera, *Gilia* has the least advanced morphological and cytological properties, is most widespread, and has the greatest number (60) of species. It is moderately resistent to desiccation, germinates during winter rains, and blooms in spring before the hot, dry summers. In contrast, *Langloisia* contains the most advanced properties, has only 4 species, and is a desert endemic. They are small and tufted, with bristly leaves and bloom in late spring and early summer. *Eriastrum*, with 14 species, is intermediate. Like *Gilia* it occurs in woodland borders, savannahs, and deserts, has an intermediate number of advanced properties, and occupies an intermediate distribution range. It has wiry stems and small, tough leaves, often with woolly hairs. It grows and blooms in the hot, dry early summer. All this suggests that *Gilia* is the oldest genus, *Langloisia* the youngest, and *Eriastrum* of intermediate age.

The geographical centre of origin of American Polemoniaceae does not therefore coincide with the present centre of species richness, but is related to a climatically variable region. The climatic shifts were within the tolerance limits of a gradually evolving taxon. This interpretation requires a study of the taxonomy, morphology, cytology, life-history, and distribution ranges of all constituent species, together with knowledge of the climatic and vegetational history. Only with this background is it possible to see the spatio-temporal unfolding of the family.

The impact of genetical factors on temperate wheats

The third example concerns species nests originated by a genetical mechanism. In evolutionary studies of crops such as wheats, a major problem is to discover what changes caused wheat to leave its area of origin, the Middle East, and spread into cooler, moister temperate regions (Broekhuizen, 1969; Wilsie, 1962), thereby also adopting another photoperiod. The geographical distribution of cultivated wheats lies outside the range of wild wheats, which still occur where the cultivated ones originated. Moreover, in other species nests, species can move within the range of their genus, but in this crop the species differ: the cultivated crops, such as wheats, occur outside the generic range of the wild wheats, but their ranges lie within that of the tribe they belong to (Hartley, 1970).

Introgressive hybridization may solve these problems. Harlan and Zohari (1966) described the ranges of wild wheats and found for each species a region where the species occurs abundantly in natural vegetation, surrounded by areas where they occur as weeds in cultivated fields. Zohari (1965, 1969) described the ranges of weedy grass species of a related genus, *Aegilops*. These species, adapted to cooler, moister conditions, are found more to the north than those of wild *Triticum* species. Zohari also estimated the distribution of certain genome types, among them the D-genome that these species share, but which is restricted to a part of the generic range of *Aegilops*. Analysis of wild wheat species and of various, dated evolutionary stages in the cultivated ones, show that *Triticum* species absorbed, apart from other genetic material, this D-genome from *Aegilops squarrosa* through introgression, which enabled them to shift and expand their ranges northward into temperate regions. There they speciated, forming secondary centres of specific concentration (Zohari, 1970).

Thus, introgression of genetic material from species from different genera of the same tribe enabled the wheats, together with other crop species, to extend and shift their original ranges. This also explains why species nests cannot be defined at a generic level, but at a tribal level. Their ranges conform to their tribal range and to the factors determining this (Hartley, 1970).

These three examples defined species nests either at the next higher taxonomic level or an even higher level. They also showed three ways species nests can originate by differences in germination physiology, phylogenetic adaptation, and genetic mechanisms, although other ways can occur. The temperate-wheat example discussed only one trait, although more are known, complicating the process even more. Apart from physiological ones, morphological changes occurred, resulting from selection or genetic introgression.

The same may be true for the causation of species nests in general: many traits at all levels of integration and spatio-temporal scales may occur. Their common denominator is adaptation to ecological conditions that vary in space and time.

Conclusions

This chapter discussed geographical gradients in species number, possibly explicable by present-day or past factors. It appears that the broadest gradients can be explained in many ways, and that it will be difficult to test them, if at all possible. Going to spatially and taxonomically more restricted species assemblages, testability increases, possibly as a result of the increase in biological insight gained. This gain is particularly important in judging whether or not to undertake biogeographical study. Analyses of global variation or of large or too heterogeneous taxa often leave us empty handed, despite the amount of work done. They show us little about continuous spatial and evolutionary adaptations that give rise to present-day distribution patterns.

The next chapter continues along this line, but instead of considering gradients in species richness irrespective of their properties, it concentrates on geographical trends in a number of species properties, still irrespective of the species' identity.

7

Geographical trends in biological traits

Historical explanations should, in general, be considered when a distribution pattern cannot be explained fully in terms of responses to contemporary environmental conditions. A species could, in theory, occur in a certain region, but may not have reached it yet. Limited dispersal capacity can cause a species to be absent even where conditions are optimal for it. Often it is difficult to determine how far actual and potential patterns of distribution agree or what is the potential range. For example, it is difficult to judge if some Eurosiberian or Australian species might be able to grow, possibly even grow well, in North America. Some species transplants show it is possible, but we also know that in many cases, perhaps even the majority, transplants fail. Another problem is that the extent to which actual and expected distributions agree depends much on spatio-temporal scales of analysis.

To judge to what extent actual species ranges reflect preferred ecological conditions, one must estimate their possible adaptation to these conditions. The more their preferences fit local conditions, the less we need to consider historical conditions to explain their distribution patterns. Such measures of adaptation to local conditions include those suggested by global trends in leaf form, plant life form, and genome size. These traits and three others are discussed in this chapter.

Many cases of adaptation to local conditions are inferred from a supposed rapid migration into localities with suitable conditions. In other cases, when their properties alter, species remain where they are. Examples of changing biological properties within a species range include differential adaptation to local climatic conditions, as has happened in the house sparrow in North America discussed in the next chapter, or in local adaptation to predation pressure, as has occurred in molluscs in different parts of the oceans.

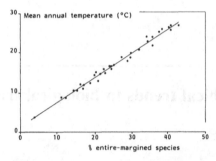

Figure 27. Proportion of the number of plant species with entire margins plotted against local mean temperature (after Wolfe, 1978).

Leaf form in plants

Information from 47 floras of various parts of the world, covering 101 536 plant species enabled Bailey and Sinnott (1916) to describe a global trend in the proportion of species with entire or non-entire margined leaves. 'Non-entire margins' comprise crenulate, crenate, serrulate, lobed, incised leaves, etc. Thus, they showed that the highest percentage of species with entire-margined leaves occur in tropical lowlands, and the lowest percentage in cold-temperate regions. Species with non-entire leaf margins, when occurring in the tropics, occur particularly in moist conditions at high altitudes, in ecologically stable conditions, and in relatively cool habitats (Figure 27). Conversely, outside the tropics species with entire leaf margins are found in arid or physiologically dry habitats. This pattern is stronger in trees than in shrubs and herbs, possibly reflecting differences in life habits. These leaf-form proportions have been used to reconstruct climatic conditions in the geological past; for example, temperature has thus been shown to decline during the Cainozoic (e.g. Wolfe, 1978).

Apart from this geographical pattern, species with single or compound leaves show a similar geographical trend as well. For example, Webb (1959) found that the percentage of species with compound leaves is 0, 10–30, 30–40, and 15–50 in cold–temperate, warm–temperate, subtropical and tropical parts of Australia, respectively. Working with dicotyledons only, Stowe and Brown (1981) found that the highest number of species with compound leaves occurs in those parts of North America where spring and summer temperatures are high, particularly where precipitation is low. This relationship between percentage of compound leaves and climatic conditions varies between families: species having small leaflets – Leguminosae and Rosaceae – generally occur in dry regions, whereas those with relatively large leaflets favour warm, humid areas.

As both trends occur in many families, it is likely that spatial adaptation to climate is the main reason for this relationship, rather than evolutionary adaptation. The actual ecological advantage of one leaf form over another is, as yet, unclear.

Life form in plants

Raunkiaer (1934) classified plant species from throughout the world into a few categories according to how they survive unfavourable periods. One category, therophytes, only survives such periods as seeds, which germinate when conditions become favourable again. Individuals of species belonging to other categories do not die off completely as therophytes do. Instead, they die off only partly, or shed their leaves, after which they bud again when conditions improve. Buds of species that only shed their leaves may be found high above the ground (phanerophytes), near or close to the soil surface (chamaephytes), or at a certain depth within the soil (hemicryptophytes).

After assigning species into these categories, Raunkiaer calculated the percentage of species of each category for various latitudes in the Northern Hemisphere. He demonstrated that the tropics contain the highest percentage of phanerophytes, and the temperate zone the highest percentage of hemicryptophytes. Chamaephytes prevail further to the north. Thus, tropical plants generally have their buds high above the ground, whereas in the cool, temperate zone buds are near or under the soil surface, and in cold, arctic areas they are found slightly above the soil surface. When a desert interrupts the latitudinal temperature gradient, the trend from phanerophytes through hemicryptophytes to chamaephytes is broken by a high percentage of therophytes. The same trend is often found along altitudinal transects in mountains, as the altitude where 10%, 20%, or 30% of the species are chamaephytes varies with latitude.

Raunkiaer related this global trend in life-form proportions to locally prevailing temperatures. The strong increase of chamaephytes at higher latitudes is, according to Raunkiaer, explained by tussocks and mats of intertwined shoots being directly warmed by the sun's rays without interference of the soil, which often remains cold or frozen for part of the year. This explanation accords with that for the coincidence of the line in the Northern Hemisphere connecting locations with a flora containing at least 20% chamaephytes with the June isotherm of 4·44°C. The line connecting a minimum of 10% chamaephytes in local floras coincides with the June isotherm of 10°C.

On the more regional scale of Greenland, Sørensen (1941) divided the flora into northern and southern species. Although the northern group did not show an excessively high percentage of chamaephytes, it was higher than that of the southern species. Moreover, buds of northern species on the whole lack protection during winter, whereas buds of southern species are protected. This accords, among other things, with an aperiodical initiation and development of individuals during the growing season, allowing them to utilize effectively the rather brief growing period. This appears to be less strong in southern species.

The relationship between life form and climate is well developed in various parts of the world and at different altitudes. It occurs despite the taxonomic heterogeneity of local floras and possibly reflects plant responses to local conditions that are probably of adaptive value, irrespective of other properties of the species. Its adaptive value may be large because selection may be different for other properties, possibly pulling the species in other directions as well. Yet, floral composition may have become adapted soon after the last glaciation, suggesting fast biotic responses. Moreover, climatic conditions have not been stable since the last glaciation, but have fluctuated considerably.

Polyploidy and genome size in plants

Much attention has been given to geographical gradients in genetical constitution within and between species. One such gradient concerns polyploidy, which in the Northern Hemisphere often increases with latitude. The polyploid form often has a wider geographical distribution than the diploid and may occur at higher altitudes and, locally, under colder conditions. Polyploidy may have direct or indirect adaptive value. This gradient not only occurs in many plant species (e.g. Gottschalk, 1976), but also in invertebrates, where it has been related to parthenogenetic reproduction. The fact that great numbers of species, or intraspecific forms, have become adapted spatially over large areas in the relatively short time since the last glaciation suggests that species or morphs within them have a relatively rapid spatial adaptation to geographically varying ecological conditions. A similarly rapid spatial adaptation can be seen in the global distribution of genome size in plants.

The distribution of genome size has only been investigated systemati-

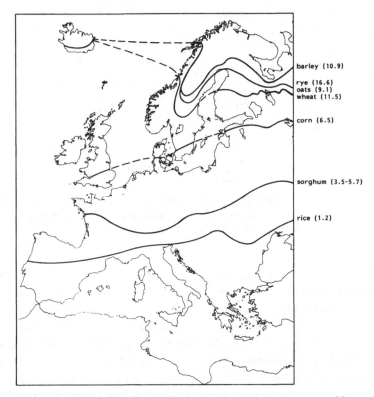

Figure 28. Relationship between genome size and latitude as expressed by northern range limits of various crops in Europe (after Bennett 1976).

cally for many species since the early 1970s. For example, in several European crops genome size is greatest in the most northern species, whereas species whose northern limit of cultivation only reaches Southern Europe have the smallest genome (Figure 28). The same relationship with latitude holds in European Russia and North America, and between genome size and altitude. Furthermore, the ranges of two of the northernmost species with large genomes, rye and oats, are not transequatorial; wheat and barley can be cultivated near the equator, but occur there only in cooler environments, namely at higher altitudes, except in summer. At the other extreme, species with small genomes, having 6·5 pg of DNA or less, are all transequatorial and have no southern limits at any season. Thus, southern limits of these species with great amounts of DNA and cultivated in summer follows more or less the same order as their northern limits in winter (Figure 29) (Bennett, 1972,

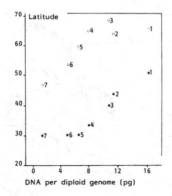

Figure 29. Relationship between genome size and latitude as expressed by the southern limits of several crops in summer (○) as well as their northern limits in winter (●) (after Bennett 1976).

1976). The same trends, though less conspicuous because of their small number of species, are found in two other groups of cultivated plants, namely pasture grasses and pulse crops (Bennett, 1976).

Levin and Funderburg (1979) found a similar positive relationship between genome size and latitude in 332 tropical and 524 temperate herbaceous angiosperms, whether subdivided into monocotyledons and dicotyledons, or not. But the relationship sometimes breaks down at the family level: Gramineae show a geographical trend in genome size, whereas Compositae, Liliaceae, and Leguminosae do not. Moreover, cosmopolitan families do not show a greater variability than geographically more restricted families. This means that families of species with different genome sizes replace each other from south to north.

These patterns have been explained by, among other things, differences in speed of cell physiological processes. Duration of developmental processes, such as cell-cycle time, meiotic duration, and minimum generation time are affected by amounts of DNA present in the cells (Bennett, 1972). For example, the life cycle of *Senecio vulgaris* and *Stellaria media* with a DNA content of 5·2 pg and 3·8 pg, respectively, takes 5 or 6 weeks, whereas *Fritillaria* species with a DNA content between 159 pg and 295 pg (420 pg) do not flower until 5 or 6 years after germination (Bennett, 1972). Clearly, genome size, and thus its global gradient is possibly of ecological significance.

Grime and Mowforth (1982), studying genome size of 169 plant species in Britain, elaborated this further. In Mediterranean species large

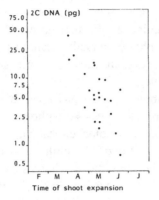

Figure 30. Relationship between time of shoot expansion of various Central English plant species during the year and their genome size (after Grime and Mowforth, 1982).

genome sizes occur in plants growing under cool conditions in winter and early spring. These species survive dry, hot summers in a dormant state as seeds or geophytes. Since low temperatures limit cell division, each species separates a period in which meiosis occurs (summer), from that of cell expansion (winter and early spring) (Figure 30). Northern European species have, according to this idea, small genome sizes, because colder winter and spring conditions delay germination, and warm conditions induce shorter growth in summer. Because phenology is possibly related to temperature from February to August, this relationship with species' phenologies is a local, temporal expression of the relationships found by Bennett (1972, 1976) and Levin and Funderburg (1979).

Thus, local and global gradients in ploidy and genome size suggest that such traits at a low integration level are probably of adaptive value to present-day environmental conditions.

Photosynthetic pathways

Recently, much has become known about the biochemistry and physiology of photosynthesis. Here geographical gradients are also apparent, particularly in the relative representation of the three photosynthetic pathways. They are known as C3, C4, and CAM (Crassulacean Acid Metabolism) photosynthesis, and reflect differences in atmospheric CO_2 fixation. Most species utilize only one of them, although in CAM species a switch from C3 to C4 photosynthesis is possible. C3 species are by far the most common and widespread, whereas C4 photosynthesis is found in a relatively small number of plant families, mainly Gramineae. Furthermore, C4 photosynthesis is generally absent from phylogenetically primitive families and is thus possibly more specialized than C3 photosynthesis (cf. Edwards and Walker, 1983, for details).

C3 species utilize ribulose biphosphate carboxylase (RuBP) to fix atmospheric CO_2, whereas C4 species use phosphoenolpyruvate carboxylase (PEP). Apart from other biochemical characteristics, CAM species can utilize both C3 and C4 pathways. C3 and C4 species photosynthesize under light conditions, whereas CAM species photosynthesize under dark (PEP) as well as light (RuBP) conditions. The functional difference between C4 and C3 photosynthesizers is that C4 plants fix CO_2 more efficiently at low intercellular CO_2 concentrations than C3 plants because of the greater affinity PEP carboxylase has for CO_2 than RuBP carboxylase. The consequences are that C4 photosynthesis is not saturated at midday light intensities, whereas C3 photosynthesis is saturated by 1/3 of that intensity. C4 species have a two-fold increase in water-use efficiency relative to C3 species, and maintain positive net photosynthetic rates at higher leaf temperatures than do most C3 species. Favoured leaf temperatures of C4 species are between 30°C and 45°C, whereas the thermal optimum for most C3 species ranges between 15–25°C. As atmospheric CO_2 uptake may occur at night in CAM species, their efficiency in water-use is much greater than in C3 and C4 species. At night the gradient of water-vapour pressure between the transpiring organ and the atmosphere is smaller than during the day, ensuring transpiration rates are lower at night.

The ecological consequence of these physiological processes is that under high temperature conditions, high light levels, and low moisture, C4 plants may have a competitive advantage over C3 species. Thus, C4 species will be most abundant in warm climates, for example tropical lowlands and deserts (e.g. Hattersley, 1983). For example, few C4 species of Cyperaceae occur in the North American arctic tundra (Figure 31a), if at all, whereas up to 82% of Californian grasses may be C4 species. Within C4 species one can distinguish monocotyledons and dicotyledons (Figure 31b), the latter being more efficient under moisture pressure at high temperatures than the first. CAM species correlate negatively with humidity; CAM succulents are most abundant in regions with dry soils. Within the succulents, one can distinguish between the physiological requirements of Crassulaceae and Cactaceae (Figures 31c and d), the former being most abundant in regions with low soil moisture

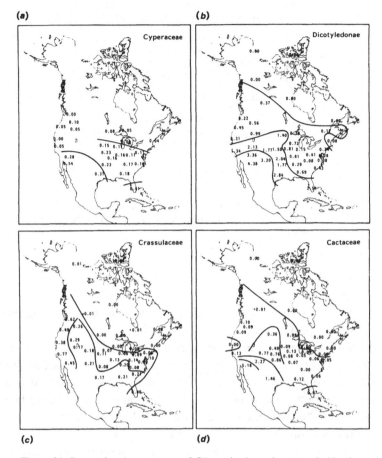

Figure 31. Proportional occurrence of C4 species in various taxa in North America (after Teeri *et al.*, 1978).

and precipitation, whereas Cactaceae are most abundant in regions with low soil moisture and high evaporation rates (Teeri, Stowe and Murawski, 1978).

Differences in geographical distribution of C3 and C4 species are paralleled on finer spatial scales, as well as in their seasonality. Firstly, the decrease of C4 species with increasing latitude also occurs as a steep altitudinal gradient (Cavagnaro, 1988; Teeri, 1979). An excess of C4 species occurs in open, dry localities, whereas C3 species are more frequent under shaded, moister conditions (cf. Wentworth, 1983). Moreover, seasonality of the two C A M families Cactaceae and Crassulaceae differs in North America, the former being restricted to the cool part

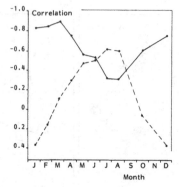

Figure 32. Correlation of relative abundance of Cactaceae (○) and Crassulaceae (●) in North American vegetations over the year (after Teeri *et al.*, 1978).

of the year, whereas growth of the second is limited to the warm season (Figure 32).

Of all the possible ecological factors, abiotic factors do not seem to be the only important ones (e.g. Chazdon, 1978). Competition as a process of biotic interactions may also play a part. For example, Teeri (1979) suggests that C3 species may be restricted to the cool part of the year through competition from the more advanced C4 species during the warmer months, but data seem to be lacking to test this hypothesis. In other cases the phenology of photosynthetic pathways operates through the regional precipitation regime (Prins, 1988).

As earlier in this chapter, we see a close correspondence between local and geographical patterns of distribution of species on the one hand, and ecologically relevant physiological properties for widely differing taxa on the other. However, as the C4 pathway is but a part of a larger suite of biochemical, physiological, and morphological traits, together forming an ecological response syndrome, its adaptive advantage and geographical significance cannot be generalized simply (Pearcy and Ehleringer, 1984).

Shell morphology in marine molluscs

The previous sections discussed global or continental gradients in ecologically important properties that many species share and that often parallel abiotic environmental conditions. Morphological gradients also exist that are best explained by biotic factors, such as global variations in predation or grazing. These gradients do not depend on dynamics in species ranges; in principle, they can be static over long periods, the adaptation being evolutionary *in situ* rather than through shifts of species ranges or dispersion of individuals within them. Of course, the ranges concerned have not always occupied the same area; just that changes in distribution patterns are not always essential to explain gradients in species' properties.

One example is the latitudinal gradient in shell morphology of marine molluscs. Vermeij (1978) found that the proportion of species with several defensive properties increases towards the equator. Snail apertures may or may not be toothed and elongated, the operculum flexible or not, and the external sculpture may vary in strength. The relatively high frequencies of snail species with the strongest defence structures in the tropics are paralleled by gradients in similar properties in bivalves, in grazing pressure in algae and barnacles, and in features of the predators, such as the crab pincers. They are explained by increasing predation and grazing towards the equator. Vermeij suggests that abiotic ecological conditions become more stable and less stressful towards the equator, and biotic interactions among species become more important in this direction.

In the tropics, the same gradients are also found locally at different intertidal levels. In contrast to the lower levels, the higher intertidal levels have the lowest frequency of armoured species. Effects of geographically temperate conditions are thus reflected by the high intertidal level and those of tropical conditions by the lower level. Vermeij provides information on mechanical action in the wave regime, on tides, temperature, and productivity to indicate that these parameters cannot account for the latitudinal variation.

Effects of the recent introduction of a predator crab, *Carcinus maenas*, on the eastern North American coast (Vermeij, 1982) suggest that prey species can adapt quickly to changes in predation pressure. One of its prey, the snail *Littorina littorea*, the larvae of which are widely dispersed, shows no effect to this newly introduced predator, even after being affected by *Carcinus maenas* for only seven years at the most. The effect on another species, *Nucella lapillus*, was marked, resulting in an increased spire height and mean adult lip thickness.

In conclusion, these data suggest that spatial trends in shell morphology need not involve geographical shifts in species ranges, as obviously occurred in terrestrial biota of the temperate region after the last glaciation.

Alkaloid-bearing species

Another example of a property possibly caused and maintained by biotic factors is the presence or absence of alkaloids as a secondary physiological product in plant cells. Not all species possess alkaloids. Those that have them, differ widely in the number of different molecules, with up to 74 in *Catharanthus roseus* (Levin, 1976). Alkaloids are found in many families: in 1970 about 300 families with 7500 species and about 4000 different molecules were known (Raffauf, 1970). They occur in plants of different longevities, ranging from annuals to trees, although annuals have a significantly higher proportion of alkaloid-bearing species (33%) than perennials (20%). Alkaloid production is genetically controlled, which makes them species specific, although environmental factors may modify their expression. They are found in all parts of the plant; however, geographical trends in the proportion of species containing them are based on their presence in leaves and seeds only (Levin, 1976).

Tropical families contain a significantly higher proportion of species with alkaloids than temperate ones. Cosmopolitan families such as Leguminosae and Compositae contain a lower proportion of alkaloid-bearing species than temperate and tropical families. A comparison of temperate and tropical species, irrespective of family, shows the same pattern, temperate areas having a lower proportion of alkaloid-containing species than the tropics. The same relationship occurs locally, species living at high altitudes contain a lower proportion of alkaloid-bearing species than those in tropical rain-forest at lower altitudes. Within rain-forests, lowland rain-forest had the highest proportions, whereas communities at higher altitudes had lower proportions.

Levin (1976) explains this pattern in alkaloid incidence in plants in terms of latitudinal variation in biotic pressure by, for example, microorganisms, insects, and mammal herbivores. Alkaloids may inhibit growth and development, or both; to mammal herbivores they may taste bitter and affect growth, development, or even kill the animal. Mammals may be able to exclude or detoxify them.

Effects of alkaloids on widely different plant predators suggest that selection may favour their development. This accords with one hypothesis concerning an increase in biotic influence towards the tropics, mentioned in Chapter 6. It also accords with alkaloid rarity in aquatic plants. Differences in pest pressure also explain why islands have fewer alkaloid-bearing species, as islands commonly have fewer predators than continental regions (Levin, 1976). One of these 'islands', Australia, in fact has relatively few large browsers, which may explain why Australian *Acacia* species have no myrmecophytic structures or spines, whereas these features occur outside Australia.

Thus, this latitudinal gradient in a physiological property may also relate to biotic factors, just as in the marine species discussed above.

Conclusions

Geographical patterns in several species traits all parallel trends in their regional or local biotic and abiotic environment. Unfortunately, there are no studies about possible time-lags, reflecting the importance of history, or rates of adaptation in these traits. It is difficult to know whether adaptation is evolutionary, spatial, or both. But as most of these gradients have been studied in temperate regions which, for large areas, were ice-covered during the glacials, spatial adaptation to local ecological factors mainly accounts for the correspondence. This conclusion concerning temperate species may also apply to several other biological properties of plants, such as photosynthetic and rooting system (Mooney and Dunn, 1970), seed size (Howe and Smallwood, 1982), gall formation (Fernandes and Price, 1988), and pollination type (Regal, 1982), and those of animals, such as insect diapause (Danilevskii, 1965), or the length of breeding seasons in birds (Wyndham, 1986). As these trends are found in large and often taxonomically heterogeneous species collections, species appear to be subject to many ecological factors affecting many different traits and operating at a large variety of scales.

8

Intraspecific trends

We have seen that both taxon density and ecological properties vary geographically, irrespective of species identity. This raises the question as to whether species themselves are also spatially structured, and if such a structure contains information that would aid understanding their spatial behaviour, continuity, and evolution.

For over a century, three intraspecific gradients in particular have been well-known for many animal species. These gradients relate, in various ways, to thermophysiology. According to Gloger (1833) individuals from more humid parts of a species' range are darker than in drier parts. In colder climates total body surface is often reduced because of relatively shorter legs, smaller ears, etc. (Allen, 1877), or a smaller surface-to-volume ratio of their bodies (Bergmann, 1847). Individuals are larger in colder climates than in warmer parts of a species range. Although these gradients concern relationships between general morphology of the individuals and temperature, they relate to physiology and temperature. Physiological work has been done, particularly on Bergmann's rule. Although these are the oldest and best known gradients, the genetical constitution of species is far from homogeneous geographically.

In this chapter I discuss species' range structure in relation to morphological and physiological gradients and to their genetical constitution. These gradients, or eco-geographical 'rules' or 'laws', are usually formulated in static terms, although on scales of thousands of years, a century, or even less, these 'laws' represent spatially dynamic, often discordant adaptations to a non-uniform, variable environment. By concentrating on the geography of qualitative aspects of range structure, I neglect their numerical build-up. Part III considers numerical and dynamic aspects of species ranges.

116

Morphological traits

Packard (1967) sampled the house sparrow (*Passer domesticus*) at 12 localities along a 1600 km transect across Illinois, Kansas, and Colorado, and used 785 adult and subadult individuals of both sexes. He studied, among other things, the relationship between colouration of various body parts and yearly precipitation. In both adult and subadult males, belly colouration correlated positively with mean annual precipitation and negatively with the auricular area of adult males. The latter correlation was not found in subadult males, nor did rump colouration of adult and subadult males correlate with precipitation. In females no relationship existed between colouration and precipitation.

Thus, in house sparrows only part of the populations is darker in areas with a higher mean yearly precipitation, as predicted from Gloger's rule, resulting in great contrasts in adult male colouration for high precipitation and small contrasts in low rainfall areas. Similar relations between temperature or precipitation and body colouration occur in insects such as bumble bees or spittlebugs (Thompson, 1988).

Other information from the same collection agrees with Allen's rule; tarsus length, as a measure of mean body weight, correlated negatively with altitude, and hence possibly with temperature. Johnston and Selander (1971) showed that tarsus length may be a measure of body size, whereas bill length, wing length, and body weight, possibly varying with behaviour, are not (McNab, 1971). Principal component analysis of the data enabled variation according to Allen's rule to be separated from that following Bergmann's rule. By analysing both sexes, possible differences in agreement between males and females to these rules could be estimated. Sixteen parts of the skeleton of 1752 individuals caught at 33 locations in North America were measured, but only 1333 individuals were used.

In both sexes all skeletal parts load positively on the first principal axis, indicating that this axis represents general body size. Next, component scores were mapped, which showed that the largest individuals of both sexes occur in the northernmost parts of North America, whereas small birds occur further south and along the coasts. These trends clearly represent Bergmann's rule and show that general body size increases with lower local temperatures. This is supported by a negative regression of component scores against mean January temperature at the sampling locations; the same regressions were not significant for mean July temperature.

In contrast to the loadings on the first axis, those of skeletal dimensions on axes 2 and 3 differ for males and females. Apart from the skull, all parts load significantly on axis 2. Sternum size is negatively related to all other parts, suggesting that, according to Allen's rule, the second axis represents thermoregulatory adaptations. Regression of the component scores of this axis agrees with this, being positive for both winter and summer temperatures.

Loadings of the males' skeletal parts on the third axis are all positive and significant; only the skull dimensions have negative loadings. Johnston and Selander (1971) suggest that this axis may represent properties related to food intake.

Interpretation of the two axes for female skeletal parts parallels that for males, insofar that in the females only the second axis represents thermoregulatory properties and those connected with food intake jointly. These two sources of variation are less distinct in females than in males. Regression of component scores is again positive for both summer and winter temperatures, as it is in males, thus reflecting Allen's rule for both male and female house sparrows in North America.

These studies thus show a distinction between effects on body dimensions according to Allen's and Bergmann's rules and, discordantly, according to sex. Moreover, coastal effects on body size are distinct from latitudinal ones. In different life-stages factors can also vary, selection sometimes operating either during migration or in the species' wintering quarters. Apart from geographical trends, Fleischer and Johnston (1982) found that in spring male house sparrows are larger than in autumn, the reverse occurring in females. Limbs and heads of males and females in spring populations were relatively smaller than those from populations during the previous autumn. This corroborates Allen's rule, even though the process is not spatial but operates over a brief time period. Rates of evolutionary adaptation may be high; the house sparrow was introduced into North America only about a hundred years ago.

The same patterns occurred in wing length of 4000 North American birds of 12 species (James, 1970). Although wing length may also reflect behavioural properties (McNab, 1971), northern birds have longer wings than southern ones from the Atlantic coast or from the Mississippi valley. Wing length correlates with local temperature and humidity, indicating morphological adaptation to local climatic conditions. However, it is not always certain whether local body dimensions are genetically fixed as is implicit in the eco-geographical rules, or that they can also represent phenotypic responses during an animal's development to local conditions (James, 1983).

Physiological traits

Apart from research on geographical gradients in morphological properties, gradients in physiological traits have also been studied (e.g. Blem, 1973), particularly on the North American house sparrow.

Kendeigh (1944, 1949, 1976) concludes that gradients concerning individual's energy metabolism and budget can primarily be explained by geographical temperature variation, and only secondarily by photoperiod. Temperature regulation, existence metabolism, reproduction, and daily energy budget all show an increase in energy use with lower temperature. Energy use is higher in winter than in summer. Since the total amount of energy required for reproduction is constant over the range, numbers of clutches per year decrease as a function of energy cost per clutch. Thus, energy shortage can explain the northern range limit: too little energy is available for reproduction after basal, survival metabolism. In southern deserts the species is limited by maximum temperatures during the day that exceed the bird's upper tolerance-limit.

Kendeigh (1969) also made interspecific comparisons, showing that Bergmann's rule holds for intrageneric variation. However, to allow extrapolation, two restrictions were made: the rule only holds for species weighing less than 1 kg, and where no other factors outweigh effects of body size on thermoregulation. Mayr (1956) also remarked that more than one selection factor may determine body size in animals (cf. also Peters, 1983).

This list of physiological processes and properties related to body size is far from exhaustive; Calder (1984), Peters (1983), and Schmidt-Nielsen (1984) discuss the matter extensively. These examples illustrate the existence of other relationships apart from body size and temperature. It is therefore understandable why body size of 10 North American mammal species relates negatively to latitude, 15 relate positively, and 22 show no relationship (McNab, 1971). Ecological factors other than physiological ones may play a part.

Whether or not McNab's (1971) interpretation, based on species interference through competition is correct (cf. Ralls and Harvey, 1985), only a minority of his species follow Bergmann's rule, while others show no correlation or even a 'wrong' one. This again illustrates that many selection factors may operate in combination, though at different scales and affecting different traits within each organism of any single species.

Population genetic variation

Ranges can also be differentiated genetically, taking two forms, geographically continuous variation and discontinuous variation. Phenetic properties vary continuously when dependent on many independent genes, and when environmental conditions to which they respond also vary continuously. But when the genes are arranged in a few discrete units, ecological properties may vary discontinuously in space. Within species ranges various genetic forms may be found under different climatic conditions (Prentice, 1986) and in different habitats. The plant species *Avena barbata* which was introduced from the Middle East into California about 250 years ago is an example (Allard *et al.*, 1978).

Avena barbata has two ecological forms that differ in the 35 loci analysed. One, the Malibu type, occurs in dry biotopes and the other, unnamed form in more humid places. Plants of the Malibu type occur in South and Central California, where the annual precipitation is less than 500 mm. The forms may occur together in polymorphic populations, or in monomorphic populations, separated by as little as 100 metres. The polymorphic and monomorphic populations differ in habitat preference, depending on compensatory moisture conditions, such as exposure, slope, and drainage. For example, a monomorphic Malibu population was found in the Klamath Mountains in northern California (annual precipitation 1000 mm) on slopes with a southern and western aspect and a shallow, sandy soil.

Apart from this geographical differentiation, dispersion within polymorphic populations agrees with their ecological requirements. Malibu plants occur on dry sites and the other form in moist ones. Although these two forms can interbreed and hybrids occasionally occur, they may coexist at very short distances (*ca.* 1 metre) and retain their genetical identity.

Spatial dispersion within local polymorphic populations is thus discontinuous. The populations are differentiated into a mosaic of discrete, genetically monomorphic groups, each adjusted to moisture conditions on a microscale. Depending on the prevalence of particular moisture conditions and on the sizes of areas with dry or humid conditions relative to each other, a population will be polymorphic, to some degree, or monomorphic. This differentiation occurs on all spatial scales, from broad, geographical scale down to the infrastructure of local populations.

Fine-scale spatial variation also occurs in other plant species, the genetical composition of which can change dramatically at intervals of about 50 cm (e.g. Antonovics, 1978). Because of their mobility, animals can seek out preferred breeding grounds or habitats, enabling them to escape from, or adapt closely to, local conditions. In principle, they may be ecologically less flexible than many plants and their spatial adjustment may be finer.

Variation in species' ecological requirements is, in fact, variation in average responses of different genetical forms. In the case of *Avena barbata* two forms vary in relation to habitat moisture conditions. Apart from variation resulting from genetic forms within a species or population (adaptability), differences in amplitude of responses (flexibility) can also be observed, this amplitude possibly also being genetically determined. The combined effects of adaptability and flexibility determines a species' ecological tolerance relative to particular environmental factors and may, to some extent, explain variation in geographical range size (Van Valen, 1965). Location of a species range can thus be thought as being determined by the average response of all genetical forms together, while range size may depend, apart from on polyploidy, for example, also on the number of genetical forms and the amplitude of the response of each individual form.

The genetical structure of species ranges concerns differences in geographical location of the individual genetical forms, just as differences between congeneric species form a species nest. The nested structure of species ranges within the geographical range of a genus or a higher-level taxon extends to the genetical structure of populations within a species range, or even within the range of their local populations.

Changes in population genetic structure

Populations or whole species may be buffered against fluctuations in their environment because of short-term and long-term shifts in population genetic structure. *Cepaea nemoralis* is a case in point. The genetic background of each individual can be determined from the number of bands on the shell and overall colouration and lip colour, even in subfossil material. Area effects, expressing the spatial stability of the genetic composition of populations, have lasted for about 6500 years (Currey and Cain, 1968). It is known in detail from various spatial levels in England, France, and in the Netherlands, and the ecological requirements of the genetic forms have been experimentally determined (Wolda, 1963).

Currey and Cain (1968) assembled material from archaeological sites in southern Britain and categorized this into pre-Iron Age, Iron Age, and

Recent. The proportion of unbanded to mid-banded shells was relatively high in shells from Iron Age sites and recent collections. However, the proportion in pre-Iron Age shells resembles present-day southern France assemblages, the snail living there at higher temperatures than Britain today. Laboratory results also indicate that the unbanded form is more resistant to higher temperatures and temperature exerts a strong selective pressure. According to independent climatic information pre-Iron Age temperatures were *ca.* 2·5°C higher than those of the period from the Late-Atlantic to the present, despite shorter-term temperature fluctuations. Iron Age climate was similar to today; its summers were relatively wet and cool, contrary to the warm pre-Iron Age summers.

Genetic variation may thus buffer populations or species as a whole against environmental instability on a broad temporal scale, and reflect the direction and variance of environmental fluctuations to some extent.

Discordant variation in man

The common denominator for all studies mentioned so far is that species or higher taxa segregate in space and time, by extinction, adaptation, or migration. Similar to other categories, species are also to some extent heterogeneous conglomerates of properties, each of which, within certain limits, are in equilibrium with environmental conditions. As different properties respond to different environmental variables, which in turn, vary differently in space and time, the geographical patterns and processes of species properties will not be congruent, but vary independently. Their variation will be geographically discordant. Independence of genes or mitochondrial DNA may be so great that they blur species or subspecies delimitation (e.g. Wilson and Brown, 1953). Discordant behaviour may thus expose the complexity of geographical processes in a detailed way.

Three types of processes leading to discordant variation can be distinguished: (1) differential tracking of environmental conditions by different properties, (2) different migration rates of these properties, and (3) random processes. All these types relate to particular scales of variation. Observations on broad spatial scales, for example, may show up different properties responding to variables only varying on these broad scales. Rates of migration, by their nature, are also scale-dependent, as are random processes. For several reasons human blood groups are ideal for a detailed picture of aspects of the genesis of intraspecific discordance, because the genetic basis of 39 independent blood groups is known for many locations, both at broad and fine geographical scales. Moreover, evidence on their history and ecological function is available. Finally, synthetic variables have been derived using multivariate techniques, such as principal components analysis, discriminant analysis, spatial autocorrelation, and simulation (see Ammerman and Cavalli-Sforza, 1984, for a survey of this work).

The historical information concerns estimates of the geographical decline of Mesolithic cultures and the rise of early farmers. From this and other information, estimates can be made of the expansion rate of early farming cultures over Europe using Skellam's (1951) version of the advancing-wave model. This model contains two components; a spatial diffusion process and local logistic population growth. Both components are relevant, the first for understanding spatial spread and the second for local changes in gene frequency ratios, as farming cultures have higher growth rates than hunting cultures. These estimates and those from simulation studies indicate a rate of spread of 1 km per year (compare Hengeveld, 1989a; Caughley, 1963; Van den Bosch *et al.*, 1990).

As variation of gene frequencies over Europe differs among genes, several multivariate techniques show their joint variation. I will mention results from principal components analysis. The first three components show clear geographical patterns, both separately and in combination in trichromic maps (Menozzi, Piazza and Cavalli-Sforza, 1978). Of all 38 alleles, the 21 HLA genes are easy to interpret. Axis 1 shows a concentric south–east to north–west trend, similar to the progression of early farming. Axis 2, showing an east–west trend, may represent tribal movements from Central Asia into Europe, and axis 3 either an expansion of Indo-Germanic people from the Black Sea region or barbarian invasions during late Roman times. These suggest migration of various demes into Europe rather than cultural diffusion. However, matters may be more complicated than this, involving, for example, selection against the HLA-B8 gene, which is intolerant to gluten present in wheat and barley (Simoons, 1981).

The world distribution of 39 genes (HLA and non-HLA genes together) show longitudinal trends, probably caused by tribal migrations, and latitudinal ones probably caused by some climatic agent(s). Migration is suggested by patterns in the discriminant functions and possible climatic effects by a close correlation between gene frequency and distance from the equator for the Northern and Southern hemispheres jointly. The centre of migration seems to be in Central and South Asia.

This example shows several traits relevant for discordant variation. (1) Depending on the scale of observation, demes may have migrated in different directions from different regions and at different times. (2) Rates of expansion will have been different, Asian tribes penetrating into Europe and moving fast, in relatively small numbers and during a short time, and early farmers at a rate of only 1 km per year. (3) Also, growth rate differed for different economic systems. (4) On a global scale climate may, in some way, be selective latitudinally, whereas food selection operates on a continental (e.g. lactose absorption; Simoons, 1978) or regional basis (e.g. gluten intolerance). Rh-variation, though causing high mortality in certain demographic stages, may be random.

Therefore, within one species, only a few properties may show considerable variation in spatio-temporal dynamics, influenced by different environmental or demographic factors operative at different times and on different scales. This variation is similar to that between species and supraspecific taxa.

Conclusions

Each species is subject to several selection factors that vary geographically in steepness and direction. As these factors not only differ for various properties within species, but also between them, Mayr (1956) stated that eco-geographical rules concern intra-specific geographical gradients only. If Mayr is correct, size relations in local communities may be approached more profitably from the geographically individualistic viewpoint, rather than from that of community characteristics.

Whatever the complexity of processes operative at one or more scales and the techniques of measurement and data processing, it is clear that environmental factors are all-important in determining this aspect of geographical variation, both in space and time. This applies to morphological, physiological, and population genetic traits at both broad and fine spatio-temporal scales.

Summary of Part II

Part I discussed biogeographical classification and ordination, and Part II gradients in taxon density and biological traits. Although these subjects seem far from each other, from a methodological viewpoint they are related. Classification and ordination concern the topography of a taxon's geographical distribution pattern. But we can also classify or ordinate biogeographical phenomena differently, for example, according to biological criteria. In Part II these biological criteria were categorized into taxon density on various spatial scales and into those concerning qualitative properties among and within species. Part III will consider numerical spatial patterns within single species ranges as another biological criterion to categorize biogeographical phenomena.

Since these criteria all concern different aspects of the biology of species, any categorization applying only one or a few of them distorts the picture of biogeographical variation as a whole. To understand species biogeography, we must integrate all information obtained in various ways.

Species, to a greater or lesser degree, conform to general patterns of geographical variation. This applies to categories containing species of the same taxon, such as genera, families, or orders, or to a taxonomically heterogeneous set of taxa. Applying a certain categorization, they seem to share ecologically relevant properties, or a common history, which often does not become apparent when another criterion is applied. Applying different biological criteria gives different information about species, just as various numerical classification and ordination criteria and procedures do. For any individual plant of whatever species in the arctic, for example, it is lethal to expose unprotected buds several metres above the ground, whereas in the lowland tropical rain-forest it has advantages.

We apply several criteria and procedures and choose several spatio-

temporal scales of variation and taxonomic levels, not because effects of
these factors thus detected are restricted to a particular scale or taxon, but
because these effects can be optimally detected that way. In principle,
environmental factors operate on all scales and taxonomic levels, but are
often differentiated on one or a few. For example, in contrast to soil-
moisture content, not finding any temperature effect on the local dispersion of, say, beetles or plants on a certain day, does not imply that
temperature does not affect them, whereas soil moisture does.
Temperature may determine the general probability of survival, while
soil moisture determines that at individual sites. Temperature effects may
be uniform on a local scale, and hence not differentiate at that scale. They
just cannot be detected by relating numbers of organisms with site-to-site
ecological conditions.

   Results, either from classification or ordination (Part I), or from
studying inter- and intraspecific gradients in qualitative properties (Part
II) all contain information relevant to understanding biogeographical
patterns and processes. Ignoring different disciplines or results of different comparisons is like putting on blinkers and thus blinding ourselves.
In Part III the same inductive approach will be applied to spatio-temporal
variations in intraspecific quantitative species properties.

III

Areography: the analysis of species ranges

Part III concerns quantitative intraspecific geographical variation and describes species ranges and their structure and dynamics. Species ranges are commonly conceived as numerically uniform and static, whereas in reality they are structured, both qualitatively and quantitatively. Moreover, when viewed over timespans of daily through seasonal to millenia scales they are highly dynamic, both internally and externally. Both range structure and its dynamics require a special look at geographical variation; we cannot use deterministic models to describe ranges, but have to consider stochastic ones. It is necessary to define geographical processes in terms of frequency distributions and shifts therein, expressed by spatial shifts in the range location. The range can split up temporally into two or more nuclei, or pulsate, contract, or extend. Consequently, range dynamics involves more than colonization; it involves changes in numerical abundance over the species range, sometimes, but not necessarily, expressed by shifts in location, or as temporary expansion or contraction.

Ranges can usefully be conceived as broad-scale response surfaces, the highest abundances in their centre indicating conditions most favourable to the species, and the gradual decrease towards the margins as a gradual worsening of these conditions. Thus, ranges conceived as response surfaces interpret abundance distributions in eco-physiological terms and optimum-response surfaces indicate these abundance patterns over the range as a response to geographically varying ecological conditions. These patterns are expected; they vary with time. Thus, for, say, n years we can count the number of realized abundance levels for each locality within the range and construct a two-dimensional frequency distribution for all locations in the range. Within this period, frequency distributions will not remain static, nor will the distribution from one period to another. Choosing a certain value of n determines the level of temporal resolution and hence the level of spatio-temporal dynamics of the species and its causes. The same holds for other population characteristics than the species' numerical abundance, necessitating a general term of species' local performance. Conforming to general terminology outside biogeography, I propose to use intensity of occurrence (Hengeveld and Haeck, 1982).

If species ranges are visualized as broad-scale optimum-response surfaces, with temporal frequency distributions and dynamics reflecting changes in the species' environment, it is assumed that this environment similarly varies. Thus, a species' environment should be characterized in terms of frequency distributions and not as uniform and static categories. Adopting this statistical approach, we can describe species ranges and processes partly in stochastic terms. One application of stochastic models is the description of spatial processes at range margins in terms of epidemic models; another in population dynamics is the 'spreading-the-risk' concept. The problem why local numbers do not expand to outbreak proportions, nor gradually decline, resulting in species' local extinction is not confined to ecology; it also applies to species ranges. Moreover, interpretation of local intensities is relevant to the geographical build-up of species ranges. Chapter 11 thus discusses population dynamic theories, even though this topic is usually confined to ecology.

We must also look the other way, that is at broad-scale dynamics of species ranges, covering thousands or millions of years. Information is now available for species dynamics during the Quaternary, resulting from climatic change, particularly since the last glaciation. This new information is of utmost importance in understanding species ranges as dynamic structures. Defining ranges as dynamic structures allows us to understand a species' biogeography; the dynamic range structure, as seen as an optimum response to a spatially ever-changing environment, is thus basic to dynamic biogeography.

9

The anatomy of species ranges

The idea underlying Parts I and II was that individual data should be compared to generate hypotheses as to the cause(s) of the patterns found. This idea also underlies the present chapter, where all local intensities within a species range are compared to generate hypotheses as to the causation of various aspects of these ranges. Ranges represent local response intensities to environmental variables, often including a response lag. They are composite reflections of the multifarious aspects of a species' biology. Similar to a species' local ecology, but even more complicated, a species' biogeography results from processes ranging from its molecular biology to its population biology, or even to aspects of its community ecology.

Extending spatial and temporal scales from the local scale to the often long-term geographical scale exhibits even more processes, i.e. those of living in quite different conditions in various parts of the range and those of other historical or geological periods. It also exhibits responses not immediately apparent at spatially broader and long-term scales such as those discussed in Parts I and II. Finally, because differences between local environmental conditions vary over time on all scales, species ranges are dynamic rather than static. To match this dynamism, species have all sorts of adaptation, such as genetic variability and adaptability, physiological flexibility, anatomical and morphological traits, behavioural properties for rapid or regular dispersal or migration, or for hibernation, and ecological ones relating to their survival probability, such as a certain reproduction potential, longevity, diapause, etc., and adaptability to different community compositions. When one or a few of such adaptations falls short, the species will die out.

This chapter, although going into detail on many aspects of species ranges, still gives only the outer appearances of this dynamism; the picture it gives is a mere snapshot of an intricate and complex process of perpetual adaptation on all spatial scales. Chapters 10 and 11 discuss the dynamics as such on geographical and local scales, respectively.

Yet, in a snapshot one can already discern some aspects of the process. Even on a short, momentary scale, species ranges show up the responses of all individuals together to their environment. Together they form a response surface as the responses occur in two-dimensional space. And as they are not linear but humped relative to increasing intensities of environmental variables, these surfaces are optimum-response surfaces. I will therefore first describe species ranges in terms of optimum-response surfaces, and then their build up, shape, and extension, and highlight their delimitation. Although ranges can be conceived of as broad-scale optimum-responses surfaces relative to abiotic factors, species interactions as biotic factors also require special attention. Finally, the application of response surface methodology to climatic reconstruction will briefly be discussed.

Range structure

Many biogeographers are interested in environmental conditions that either determine location of ranges as a whole, or in conditions at range margins, whereas ecologists are usually interested in local conditions and processes within species ranges. Consequently, ecologists study the effects of local conditions on population parameters and neglect the geographical context. Biogeographers, on their part, do not usually consider within-range local conditions. They may fit isopleths of climatic elements, such as mean January or July temperature, to range margins. Both interests can usefully be combined, because species performances vary within ranges, due to spatial variations in ecological conditions and population processes following different courses or varying in intensity. This spatial variation is referred to here as the structure of a species' range and comprises several, different phenomena. They all express the range as a response surface relative to abiotic ecological variables operating at a geographical scale.

The range as an optimum-response surface

Density as a species' intensity parameter is not uniformly distributed over its range, although most population dynamic models assume it is. Hengeveld and Haeck (1982) showed that, statistically, species intensities decrease from the range centre to the margins. They assigned ranges of 369 Dutch ground-beetle species, 1062 Dutch plant species, 192 species of Dutch breeding birds, 818 plant species from

Warwickshire, and 211 species of Dutch breeding birds from Great Britain and Ireland to four categories. These categories were species found at or near the range margin, and near or at the range centre. They assigned these species to logarithmic classes of local frequency of occurrence as an intensity measure. From counts of numbers of species per frequency class in each category it appeared that for all taxa central frequencies are statistically higher than marginal ones, the mean frequency gradually decreasing from the centres. This is also found in unselected species sampled in different sized areas, suggesting that it is general for all species and is scale-independent (Figure 33). The optimum surface distribution also accords with one of the possible models explaining the general species-abundance relationship, the surface area – and hence the probability of discovery – of that part of the range with the lowest abundances being higher than that of the core area with the highest abundances (Hengeveld, Kooijman and Taillie, 1979).

Abundance level also relates positively to numbers of locations in which a species locally occurs (Figure 34) (Hanski, 1982), although numbers of habitats in which a species occurs, decrease from range centre to margin. This wider habitat choice in the range centre than in the margin is also reflected by central species categories found in syntaxonomically higher vegetational units and marginal ones in lower units, respectively. Finally, plant populations are more disperse towards range margins and more aggregated in the centres (Hengeveld and Haeck, 1981; Woodson, 1964). However, species are less dispersive in their range margins than in the central parts (Haeck and Hengeveld, 1977). Therefore, intensity components other than numerical abundance, such as number of favourite habitats and frequency of occurrence, also decrease from the range centre to the margin (cf. also Caughley *et al.*, 1988).

This humped distribution of various intensity parameters can be explained by assuming that, within certain limits, species responses are constant throughout the range, but that the environment varies gradually in all directions. The regularity of their geographical pattern causes species' broad-scale optimum surfaces.

Drude (1876) first described the ecological similarity of individuals from various parts of a species range which has since been described as the Law or Rule of Replaceability of Ecological Valence (Warnecke, 1936), of Relative Habitat Constancy (Walter and Walter, 1953), or of Regional Shift in Soil Stratum (Figure 35) (Ghilarov, 1959). These rules stress that species are the same throughout their range, and that they

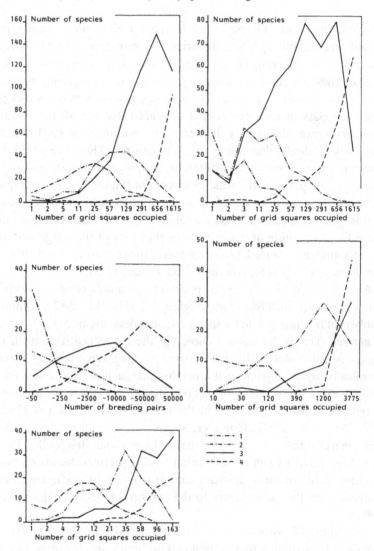

Figure 33. Frequency of occurrence of birds, beetles, and plants in various parts of Europe. The species were categorized from marginal occurrence in the sampling area (1) to central occurrence (4) (after Hengeveld and Haeck, 1982).

compensate for differences in living conditions by shifting from one preferred habitat type to another. It is also well-known that a species' altitudinal occurrence relates negatively to latitude, locally resulting in southern species occupying lower altitudes than northern species (cf.

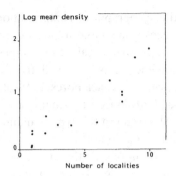

Figure 34. Relationship between number of localities of bumblebees and the logarithm of their mean density (after Hanski, 1982).

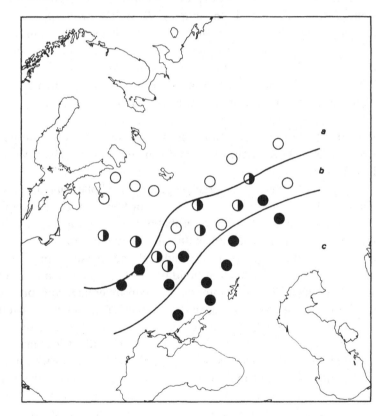

Figure 35. Shift in habitat occupancy of the beetle *Melolontha hippocastani* from open land in the northern range part (*a*) to forest in the southern part (*c*) (after Ghilarov, 1959).

Kless, 1961). Further geographical compensation concerns the occurrence of southern species on southern hill sides in the northern parts of
their range. They occupy, so to speak, 'those slopes looking towards their
home'. This phenomenon was observed by Kless (1961) in several
hundred beetle species in the Black Forest in Western Germany and in 94
species of grassland plants in Central England, for example (Hengeveld
and Haeck, 1981). The result of this type of habitat compensation is that
areas on north-facing slopes in north-central New Jersey resemble those
'300 miles further north more than that of the south facing slopes on the
other side of the ridge' (Billings, 1952). Similarly, a species' seasonality
can be shortened or shifted. Kless (1961) observed that northern species
of beetle in the Black Forest prevail in spring and summer, and the
southern ones in August. In summer, *Drosophila robusta* individuals
even differ morphologically from those from other seasons, shifting to its
geographically southern morphology (cf. also Fleischer and Johnston
(1982) for the house sparrow). At a finer scale, the same phenomenon can
be recognized: in Southern Germany, northern European and Russian
plant species avoid dry, sunny localities in Southern Germany, living in
relatively moist habitats of moors and woods, or in humus-rich localities.
In contrast, steppe species from south-eastern and eastern Europe and
Russia prefer more open, drier, and sunnier habitats (Hengeveld and
Haeck, 1981).

Habitat compensation is largely a geographical phenomenon; species
adapt to local ecological conditions according to their geographical
distribution. Both reflect basic, probably physiological properties. All
these compensations relate to spatial and temporal shifts in the species'
occurrence optimizing the fit of ecological and, possibly, physiological
properties and environmental conditions. They also compensate for
spatial or temporal shifts in intensity of environmental factors or processes through qualitative shifts or shifts in numerical abundance. Neverthe-
less, they all eventually fall short, causing a gradual lowering of species
intensity of occurrence towards the range margin and setting limits to the
distribution range. Changes in geographical intensity distributions of
environmental factors and processes cause shifts in species' intensity of
occurrence. Depending on (1) the species' sensitivity to certain
environmental conditions, (2) the geographical intensity distribution of
these conditions, and (3) their temporal frequency distribution in each
locality, geographical intensity distributions of all different shapes can
occur. They may be symmetrical and unimodal; in other cases they may
be skewed and multimodal.

The distribution of vitality and dynamic behaviour

Another intensity component of species may vary qualitatively, which, in turn, in some cases, can affect numerical representation. Individuals may be less vital at range margins than in range centres, vitality being expressed by, for example, the proportion of vegetative to sexual reproduction for survival. Kendeigh (1976), for example, proposed that birds only reproduce when surplus energy is available above that used for their basal metabolism. In marginal populations this surplus energy may not be available, forcing species to postpone reproduction or reproduce asexually, if the latter is possible.

Asexual reproduction occurs more frequently in plants than in animals, although some forms are known in the latter. Iversen (1944), for example, showed that in Scandinavia under marginal conditions ivy (*Hedera helix*) does not flower, whereas it flowers commonly towards its range centre. This phenomenon is so common that it is known as Baker's Law after his first thematic descriptions (e.g. Baker, 1959). When common, one can expect various explanations. Thus, not only may available surplus energy be important, but also low population density may be responsible (Emlen, 1973), or differences in photoperiodicity. In long-day species, for example, sexual reproduction is limited by short days at their southern range limits, and in short-day species by long days in the north (Salisbury, 1942).

Environmental favourability varies temporally, obscuring the area of asexual reproduction along range margins. Because of year-to-year climatic variability, for example, the limit of sexual reproduction in pine shifts considerably in northern Finland. Pigott (1974) observed similar year-to-year variation in fruit production and numbers of established seedlings at the northern edge of range of *Cirsium acaule* in England.

A species' vitality thus reflects environmental conditions directly, just as its numerical abundance does. Together with the degree of aggregation and number of habitats a species occupies, these properties are components of its intensity of occurrence. Although very important, numerical abundance is just one parameter to estimate a species' geographical intensity distribution. With high intensities in the range centres a species is vital or vigorous and with low ones at the margins it is ecologically more vulnerable.

European weeds could extend their ranges from the Middle East or south-western Europe into areas opened up by agriculture (Holzner, 1978). In these areas they behave as marginal species, occupying open habitats without much competition from other plants. In their range centre they integrate with the natural vegetation and are also weeds, whereas in the newly acquired parts of their range they only occur as weeds and ruderals and do not integrate with natural vegetation. Moreover, the greater the distance from their range centre, the narrower their range of ecological tolerance. This implies for this vegetation type that the greater this distance is, the poorer in species it becomes, a trend which also occurs towards higher altitudinal levels in the Alps. Yet, they maintain their ecological preferences, possibly causing their geographical range location in the south. Originally Middle Eastern species thus become more calcicole away from their range centres, whereas they occur on more soil types in their range centre. They are thermophilous species in ruderal and agrestial vegetation. The weeds with a Subatlantic or an Atlantic origin occurring in both ruderal and agrestial vegetations are mainly ruderal in Europe, where they are more confined to very acid and poor soils than to sands and granites of the areas of origin. The seasonal occurrence of both groups also reflect the conditions in their area of origin.

Present changes in farming practices have profound and complex effects on the composition of these communities. These changes cause contraction of ranges from Europe, and its plants are becoming more confined to the central parts of their range again, where they are most vigorous. In this latter aspect they vary; some species become commoner, possibly because of the retreat of other ones. Thus, lessening the general pressure or stress that interspecific competition exerts on species at the time the natural vegetation of Europe was turned, made these species weeds. This smaller biotic pressure may have allowed several species with greatest colonizing abilities to expand their ranges. Now that conditions change again, this time to their disadvantage, the species either contract their ranges, or they extend them even further than before because their former companion weeds are retreating. In this, they conform to the optimum-surface-response model, showing their individuality in the distances they penetrated into and retreat from Europe.

The extinction probability also decreases as a function of the number of populations in an area (Chapter 10), which decreases from the centre to the margins. This is more complicated, however. Analysis of the ecologically broad Dutch flora shows that species from oligotrophic habitats are more vulnerable than those from eutrophic ones but that the less threatened eutrophic habitats contain most species whose range centre coincides with the Netherlands (Haeck and Hengeveld, 1980). The extinction probability is only an extreme case of the variance of a

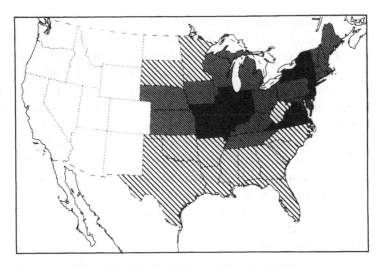

Figure 36. Number of outbreaks of the noctuid moth *Cirphis punctata* per state in the U.S. during the period 1860–1925 (after Cook, 1929).

population's fluctuation pattern; higher extinction probabilities towards the margins may imply that this variance increases in that direction. Klages (e.g. 1942) found this increase in corn, oat, wheat, barley, and rye yields for a 37-year period (1891–1927). Corn, for example, gives the highest average yield with lowest variation in the centre of the Corn Belt, the first measure decreasing and the second increasing in all directions from this centre (see also Parry, 1978; Parry, Carter and Konijn, 1988; Yoshino *et al.*, 1988, for altitudinal and latitudinal variation).

Higher average yields with smaller year-to-year variability are favourable for crop production, but not for pest species. Here too, outbreak frequency is highest in the range centre, and decreases towards the margins (Figure 36) (Hengeveld and Haeck, 1981) (cf. Rogers and Randolph (1986) for its analysis). This higher outbreak incidence in the range centre has long been known by applied entomologists (cf. Schwerdtfeger, 1968), who refer to this central area as the 'Permanenzgebiet'. The area at the margins with the lowest incidence of outbreaks is known as the 'Latenzgebiet', and the intermediate area is the 'Gradationsgebiet'. This higher incidence primarily reflects ecological conditions prevailing in various parts of a species range; neither generation number per year, which usually increases towards lower latitudes, nor the innate capacity of increase bear any relationship to it (Birch *et al.*, 1963). Salisbury (1926) distinguished three south-to-north temperature zones within species ranges: (1) a central area, where the species attains maximum density, (2) a zone where it is relatively rare and where it normally reproduces vegetatively and only occasionally generatively, and (3) the zone of cultivation where it is maintained only by artificial means. In the centre, temperature is adequate for flower and fruit production, whereas in the zone of vegetative reproduction it falls short, except in occasional hot summers. Species' survival probability beyond its natural range margin as introduced species, garden plants, or transplants may be determined by cell physiological processes, whereas in the vegetative zone the basal metabolism is just maintained (cf. also Kendeigh, 1976, for the North American house sparrow).

Cooper (1970) stated that crop production depends on conversion of solar energy and nutrients into biomass, which introduces the idea of a plant as, among other things, an energy-converting unit. As such, it requires a certain amount of energy for various functions in its life-cycle. If the amount of energy required falls short, one or more functions cannot be performed and the species may survive by performing only part of these functions or, if not, it dies out. Thus, energy shortage at its northern fringe in England may delay various stages in the life-cycle of *Cory-nephorus canescens* independently (Marshall, 1968), and together make its completion impossible within a year. The species may complete it the next year, as many perennials do in the arctic, or may fail to complete it because of many successive delays in parts of the species' margin. Its intensity of occurrence, estimated by parameters such as local abundance, would be determined by the ratio of incoming and outgoing energy in the form of germination, vegetative growth, numbers of flowers, and fruits. It may thus be useful to think in terms of energy units, such as heat units (Wang, 1963), or number of degree days, rather than in terms of average temperature or minimum and maximum temperatures (cf. Robertson, 1973). Although these latter measures may correlate with the number or energy units required by the organism, they do not give a full insight into the causal mechanism. This mechanism may explain why apparently small differences in average temperature may, in reality, prove to be large, because it does not 'correct' for the duration of the energy flow. Applying an improper correction procedure for mean temperature may result in inappropriate measures for the mechanisms concerned. Differences in average temperature across range margins or between glacials and interglacials as small as a few degrees Celsius are thus not as worrying as they are for some.

This model is not restricted to range margins only, but applies anywhere within a range. At the margins it particularly describes the relatively high extinction probability, and for the more central parts it describes the increasing probability of outbreak incidence, hence the interest of epidemiologists. From a biological viewpoint, there is no reason to look exclusively at extinction processes or at those leading to outbreaks or epidemics. The process as such is what matters.

Thus, one can study species ranges profitably from the viewpoint of changes in values of these parameters from the margin to the centre, which may show different processes operating in various parts of a range. In one part, say, high mortality rates reduce population densities, whereas in other parts effective transfer rates of propagules or individuals from one susceptible site to another reduce density. But sometimes certain parts may become unsuitable because dispersal capacity determining transfer efficiency between sites cannot compensate the frequency of unfavourable conditions. The generalization to all parts of a species range of the epidemiological model that Carter and Prince (1981) applied to range margins, gives relevance to trends in variables determining the species' range structure. These trends also make other range parameters understandable such as location, size, shape and dynamics, as well as continuous spatial adaptation and eventual evolution.

The model used (Carter and Prince, 1981) is simple; in epidemiology and population dynamics more elaborate models are used, but it contains the essential parameters; other parameters are additional to these. It contains five parameters, the number of occupied (y), and unoccupied, susceptible sites (x), the extinction rate (α), the effective transfer rate (β), and the number of propagules dispersed from each occupied site (λ). Carter and Prince (1981) were particularly interested in a minimum threshold value and hence in range delimitation; lower values would lead to population extinction or to the failure of starting a new one. Only populations, old and new, with higher values will persist. The threshold value for existing populations which do not depend on migration of individuals or on propagules from elsewhere is defined by $y_t > y_{t-1}$ and that for populations depending on a continuous flow of immigrants by $x \geq \alpha/(\lambda\,\beta)$. In the latter case, population maintenance or persistence depends therefore on a relatively low extinction rate α, on relatively high numbers of propagules or immigrating individuals with a high effective transfer rate, or on a combination of them. Two processes are relevant in studies of range margins, those leading to repeated extinction outside the range, contrasted to those of population maintenance inside. The range limit occurs between locations where one or more parameters allow the species to persist and those where they cause it to die out.

These examples refer to frequencies of occurrence for several years and, depending on annual variation in environmental factors and on demographic parameters, varying patterns of multimodal response surfaces occur. But the general trend remains the same; broadly, the highest intensities occur in the range centre and the lowest ones close to the margins. Both species vulnerability at range margins and their vigour in their centres thus point to the same interpretation, namely that species' intensity of occurrence is not uniformly distributed over the range. This intensity is highest at the centre and decreases towards the margins. This pattern is statistically defined for a particular time period. It can vary from year to year, depending on environmental conditions, as well as on species' demographic parameters.

Latitudinal and altitudinal intensity distributions

Altitudinal variations of species often reflect their latitudinal variations, where they are also sandwiched in between certain altitudinal levels and show an unimodal frequency distribution. Two aspects are clear: frequencies show a regular altitudinal build-up, and this build-up is often positively or negatively skewed (Figure 37) (Hengeveld, 1985a). However, for several reasons one cannot equate these distributions with geographical optimum-response surfaces. For example, altitudinal variation is primarily controlled by temperature along steep altitudinal, one-dimensional gradients. Geographical variation in species ranges is often influenced by more than one environmental factor at the same or different scales and their intensities often vary irregularly over two-dimensional space. Also, because species' altitudinal range is relatively small, small dispersal distances are involved, differences in photoperiod are absent, and there is relatively little opportunity for local population genetic adaptation. Furthermore, mountains give little opportunity for survival when species require extensive areas for finding food or for maintaining minimal population size with minimal genetic variation. Latitudinal environmental variation usually covers much greater spatial areas. Finally, altitudinal temporal variation in living conditions affects species more easily than latitudinal variation, because spatial dimensions on mountains can reach critical minimum levels sooner. All these differences together, in turn, affect migration by island hopping from one mountain to the other, the sieve for one species being finer than for others.

Thus, the latitudinal structure and dynamics of species ranges cannot

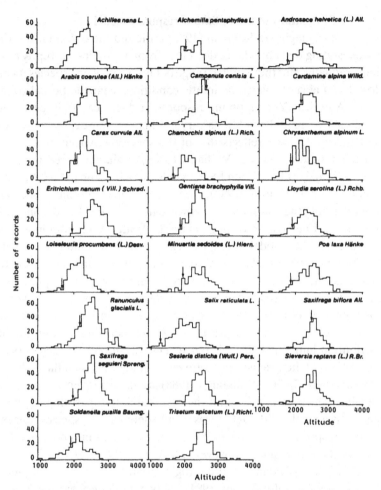

Figure 37. Number of records of 23 alpine plant species in Switzerland
obtained from herbarium material (after Hengeveld, 1985*a*, from data in
Backhuys, 1968).

be compared with their altitudinal behaviour. Such comparisons can give
some idea of, for example, mechanisms determining local differences in
the species' average intensity level, or the extent to which species respond
individualistically to varying environmental conditions. But there is a risk
in looking at species' variation at different spatio-temporal scales, as
these scales may only be partly comparable.

Evaluations of the optimum-response model

Kavanagh and Kellman (1986) criticized the optimum-response model, stating that abundance decreases from high values abruptly rather than gradually. But they studied range delimitation of one species, beech, close to an abrupt change in abiotic conditions near the boreal zone in North America. Yet, optimum-response surfaces can only be gradual when environmental conditions vary gradually; their degree of smoothness and gradualness reflects that of the species' environment.

Rather than criticizing, Williams (1988) evaluated three models to explain local densities of bumblebees in Kent: a random model, the core–satellite model (Hanski, 1982), and the marginal mosaic model as a submodel of the optimum-response one, specifically dealing with geographical behaviour at range margins. Both the random model and the core–satellite model attempt to explain hollow, bimodal frequency distributions, the first of species' spatial occurrences and the second that over time. Williams (1988) argues that neither of them applies to his bumblebees, as these are not randomly distributed but relate to vegetation type as an environmental condition or representing such conditions. Also, species seem to be present at lower densities nearer to their distribution limits than at their range centres. Moreover, species occurring in Kent at their range centre are more ubiquitous than those that are marginal there. In fact, Williams's (1988) arguments reflect the confusion – against which he fights – that the core–satellite bipartition applies to species as a whole, although it was locally defined. For species ranges as optimal-response surfaces, local abundance levels cannot be generalized to all parts of the range, however. Moreover, the positive relationship between local abundance and the regional number of sites (frequency) (Hanski, 1982; cf. also Brown, 1984) can be tautological as seen from the optimal-response surface model; frequencies of occurrence are, in fact, often used as an intensity measure at low, statistically unreliable intensities and abundance when the intensities are high.

Thus, relative to the two models assuming random occurrence, the optimum-response model explains the local occurrences of bumblebees best. The mechanism of this response would be through the animal's energy economics, which is optimal at the range centre and depauperate towards the margins.

Range shape

Internal range structure is externally expressed by the pattern of range shapes. Species ranges vary in shape, some are more or less circular, others are elongated, sometimes stretching over all Siberia and North America, but being narrow longitudinally. Such elongated ranges may indicate a narrow ecological tolerance range for one factor and a wide one for another. For example, latitudinal extension of species ranges may be limited more than the longitudinal one by photoperiod (Sheldeshova, 1967), as migration into belts with different day lengths is limited by the species' physiology and life-cycle parameters that influence photoperiodic features such as diapause. Other explanations may be that the steepness of environmental (climatic?) latitudinal gradients varies over time and that at higher latitudes they are steeper and their variation greater than at lower latitudes. Thus, the average latitudinal belt of 386 eastern North American plant species increases only slightly with increasing latitude, contrary to average longitudinal range which increases greatly (Lutz, 1921, 1922). Plants found in Colorado with different latitudinal means show the same pattern. In both cases the more northern ranges are more elongated longitudinally than the more southern ones. Of course, this pattern can only be found in regions without geomorphological obstructions or limitations. The same phenomenon, when studied in a latitudinally tapering area like South America, with, on top of this, the Andean mountain chain over its entire length (Rapoport, 1982), cannot give the same results. There the main factors controlling range shape are related to the geomorphology of the continent rather than to any biological feature.

Therefore, when topographical factors allow species to vary freely, as in North America, physiological and life-cycle properties related to species' ecology not only seem to give internal structure to a species' range, but also determine its external shape.

Range size

Species often have small ranges, only a few species have large ones, the bulk having ranges of intermediate size. Willis (1922) interpreted this variability in terms of species' age; all other things being equal, younger species would have small ranges and the older species large ones. A recent version of this theory is that of species spilling over from areas of high abundance through diffusion (Schoener, 1987; cf. also MacArthur, 1972). Yet, other things are usually not equal, but may, in reality, be the very factors that account for the variation observed.

Because range size varies continually (Figure 38), it is hard to distinguish between endemic species and 'wides', or, within the endemics, between neoendemics and palaeoendemics. By definition, neoendemics

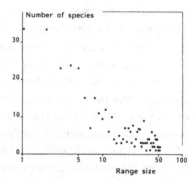

Figure 38. Number of butterfly species of various range sizes in Europe (after information in Higgins and Riley, 1970).

represent young species that would not have had time to spread and palaeoendemics old ones that have not yet died out. Palaeoendemics as remnants of formerly widely distributed species would be ecologically specialized, and have little ecological plasticity and dispersal power. This may sometimes be true, but often problems arise that cannot be solved because of lack of information. Coope (1975), for example, has shown that insect species presently found as narrow endemics in Siberia near Nova Zembla, or on mountain tops in Tibet were common 100000 years ago in the British Isles (Figure 39). It is not known if their ranges have since contracted, if they shifted location, or both. Are these species more specialized now than in the past, have they lost their dispersal power, or do they, at present, represent a phylogenetically ancient taxon? Would we have considered them specialized, lacking dispersal power or be ancient when they occurred in Britain 100000 years ago? Or do their distributions, past and present, reflect the spatio-temporal frequency distribution of a particular combination of environmental conditions that once prevailed in Britain and that at present occurs near Nova Zembla or in Tibet? Distinguishing between geographically restricted species and widespread ones, or between old and young ones within restricted species is sometimes valuable, but it often neglects ecological and dispersal parameters. Very widespread species, such as in the genus *Carex*, often have seemingly limited dispersal capacities as well. Conversely, great dispersal power may allow species to be stenoecious, enabling them to avoid or escape unfavourable conditions. Species with a limited dispersal capacity may be more flexible ecologically, although biological matters are too complex to make general statements. Often we know little about

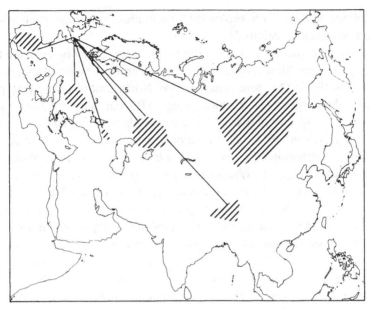

Figure 39. Present ranges of coleopteran species living together 100000 years ago at the site of Trafalgar Square (after Taylor and Taylor, 1983, from data by Coope) (1 = *Cathormiocerus curviscapus*, 2 = *Otiorhynchus mandibularis*, 3 = *Oxytelus gibbulus*, 4 = *Aphodius costalis*, 5 = *Aphodius holdereri*, and 6 = *Helophorus aspericollis*).

dispersal capacity and ecological specialization, tolerance range, or demography of many species to evaluate the factors or processes determining range size. We also know too little about past histories of species to say if they were once more widespread than at present, and if changes in their geographical extension could reflect fluctuations in ecological conditions. What we do know is that climate has always fluctuated, particularly so during the last few million years. Consequently, species or vegetation types that today are continuous over vast continental areas or different continents, were once restricted to mountain tops, and, conversely, that those now restricted to mountain tops once had a much wider distribution (e.g. Patterson, 1984; Ryrholm, 1988; see, however, also Davis and Dunford, 1987, and Davis, Dunford and Lomolino, 1988). Another consequence is that many modern vegetation types have no analogues in the not-too-distant past. Presently co-occurring species may not have co-evolved in the past, but retained their individuality, expressed in various ways. We need to supplement historical speculation with ecological observation and experimentation to answer such questions.

As an explanation, wide ecological tolerance may be expressed by a large geographical range, and narrow tolerances by small ranges (Van Valen, 1965). The more variable of two North American *Lupinus* species has the largest geographical range (Babbel and Selander, 1974). Eurytopic species of marine bivalves have larger ranges than stenotopic ones (Jackson, 1974). This relationship is also found in benthic deep-water invertebrates (Vinogradova, 1959), as eurybathic species tend to have panoceanic distributions and stenobathic species very restricted ones. We can wonder if relationships exist between evolutionary rate, or the degree to which species fragment into genetically distinct morphs, and range size. The number of subspecies or types within a polytypic species would thus relate to some measure of evolutionary rate or to range size (e.g. Rapoport, 1982). Although this approach may be obvious, the definition of subspecies as spatial units even hampers it as seventyfive percent of the individuals must differ between two popula-tions – the spatial units – to warrant a subspecies to be recognized (e.g. Amadon, 1949). A better criterion of species variability is its electro-phoretically determined genetic diversity, although usually the biological significance of this variability is unknown. Yet, populations of wide-spread species are, on this criterion, often genetically more diverse than those of restricted species.

We can also relate range size to species' local frequency of occurrence. Hengeveld and Hogeweg (1979) showed that ground beetle species with large European ranges may either be frequent or rare in the Netherlands, whereas those with small European ranges are always rare (Figure 40) (see Bock and Ricklefs, 1983; Bock, 1984). The asymmetry of this relationship between range size and species' local intensity may reflect specific ecological requirements. The larger the range, the more habitats or locations may fulfil a species' ecological requirements in its range centre, hence their greater local frequency of occurrence. Contrary to this, species with small ranges may only find a few favourable locations, not only at their range margins, but also in their range centres; ecological conditions are always marginal for them, even in the central parts of their range. Endemic species do not constitute a distinct category different from that of the wides, only the degree to which their ecological requirements match ecological conditions differs. The pattern Hengeveld and Hogeweg (1979) described, reflects the probability that during a certain period a combination of several environmental factors are either

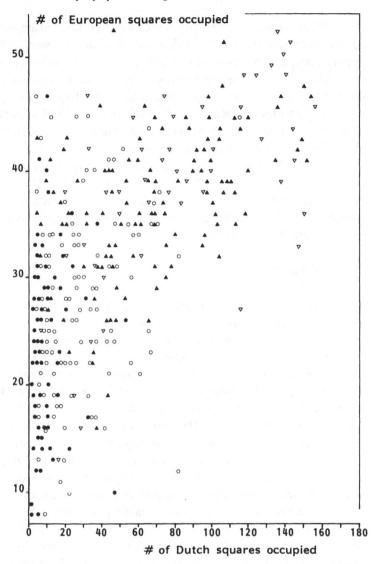

Figure 40. Relationship between the number of squares occupied by ground beetles in the Netherlands and the number of squares they occupy in Europe (after Hengeveld and Haeck, 1981).

favourable or unfavourable to a particular species, temporarily resulting in a large or small range.

Recently, various authors have attempted to explain range size in yet other terms: Hanski (1982) and Brown (1984) in terms of niches, Glazier

(1980, 1988) in those of life history, dispersal, litter size, and amplitude of fluctuation, and Bowers (1988) in those of colonization power, ecological opportunism, and competitive strength. Gaston and Lawton (1988*a, b*) emphasize the interrelationships between various characteristics. Part of these studies concern a few species only and were based on very short time series and local observation. Particularly because of unwarranted generalization from local observations to the whole range, such interpretations should be substantiated using larger data sets.

Therefore, differences in range size can be interpreted both historically and ecologically. Yet, it may be most practical to look first for ecological explanations, as they are easiest to test. Only when they fail, should we look for historical explanations, as their value depends on their weakest part, their testability.

Range margins
Indirect approaches to range delimitation
Maps usually depict ranges as solid patches or by a line indicating the range margin. Thus, they neither depict the range structure within these margins, nor the structure of the margins themselves. They are silhouettes of life happening in space.

Indirect approaches explain range delimitation by correlating or fitting isopleths of possibly causative variables with mapped range margins without using field or experimental data (e.g. Dahl, 1951). This can be done in a positive way where the range limit encompasses the area the species occupies (e.g. Dahl 1951). In the negative way the range margin is defined by factors prohibitive to the species' occurrence (e.g. Van Beurden 1981). But it is insufficient to analyse a species' geographical ecology by fitting isopleths of a supposedly limiting factor to range margins. Such isopleths, supposedly explaining the topography of part of the margin, say its northern part, concern a particular ecological variable, which does not necessarily explain the southern or eastern parts of this margin as well. These parts may remain unexplained. Moreover, the fit between isopleths and parts of range margins is often generalized, following occasional observations and average local conditions. Consequently, range margins may run circuitously if they include locations of isolated populations, peripheral to where the species' dispersion is more continuous. Finally, the choice of isopleth is arbitrary, implying that the variable chosen is not necessarily causal. For example, to explain the eastern range margin of *Ilex aquifolium* in Europe, Walter and Straka (1970) chose two isopleths: the 0°C January isotherm and the isotherm of

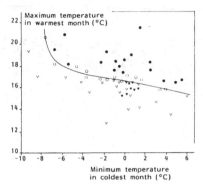

Figure 41. Plot of locations within the range of *Viscum album* (●), at its range margin (○), and outside (▽) its range in the coordinate space of minimum temperature of the coldest month and maximum temperature of the warmest. Closed triangles indicate locations with subfossil remains suggesting a different course of the range margin under warmer conditions of the Hypsithermal (after Iversen, 1944).

maximum temperatures exceeding 0°C for more than 345 days in a year. For the same margin, Haeupler (1974) notes that at least 21 other isopleths show a similar fit! Also, the coincidence of the 0°C January isotherm with this part of the range margin does not imply that *Ilex aquifolium* cannot tolerate lower temperatures for some time. The fact that it can, was shown in the 1928–9 winter when frost damage only occurred at temperatures lower than −21°C. Another example is maize in Europe. It was formerly restricted to the Mediterranean, and temperature was thought to limit its range in the north. But after breaking down the genetic component responsible for its photoperiod, maize is now a reliable crop in western Europe as far north as Stockholm.

Iversen (1944) and Hintikka (1963) improved this approach of fitting isopleths by estimating the minimum temperature of the coldest month and the maximum temperature of the warmest month for many locations inside and outside a species' range. These values were plotted graphically, the values at locations within the range being labelled differently from those outside (Figures 41 and 42). To indicate the limiting values at the species' range margin for these variables a line is drawn that optimally discriminates the two sets of points. Finally, the geographical locations of points along this line are mapped to form an isopleth (Figure 43). The correspondence of this isopleth of the yearly temperature range with the range margin provides a hypothesis to be tested by laboratory and field

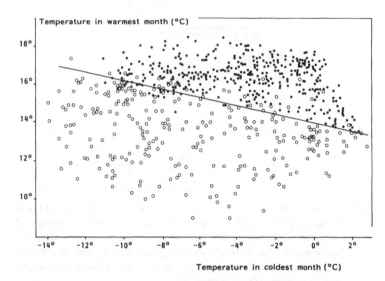

Figure 42. The same as Figure 41 for *Tilia cordata* in Fennoscandia. Dots indicate locations within the range, circles those outside (after Hintikka, 1963).

experiments; as in *Ilex aquifolium* the choice is arbitrary. The improvement of their approach is that two or more (Hintikka, 1963) variables are considered simultaneously and their analysis can be linked to the multivariate technique of discriminant analysis. Even more important is that the straight lines in Figure 42 can be related to the species' physiology, making the margin analysed understandable and new ones predictable (Skre, 1979).

Thus, one can either compare several species among each other, or compare the relationship of one species under divergent conditions at different times. In the latter case, new, hypothetical range margins can be constructed by shifting the dividing line in the graphical plot mentioned above. For example, fossils from a certain period lying systematically above this line suggest climatic change. Thus, the outlying points of Sub-Boreal finds of *Viscum album* suggest that summer temperature was 2°C lower and 1°C in winter (Figure 41) (Iversen, 1944). To test this hypothesis, Hintikka (1963) shifted the dividing line for *Carex pseudocyperus* in Fennoscandia accordingly and mapped the expected historical margin (Figure 44). Following expectation, all fossil finds from the Sub-Boreal fell within the proposed margin.

The dividing lines mentioned can either be straight or bent. This

Figure 43. Geographical range margin of *Tilia cordata* translated from the
straight line in Figure 42 (after Hintikka, 1963).

difference may either be due to differences between the species concerned, or to the spatial scale of observation: Iversen's observations
cover Denmark only, and Hintikka's Fennoscandia. If the scale of
observation is important, the difference in curvature of the dividing lines
suggests that a species' response to a particular combination of
environmental variables varies over its range. This variation may reflect
geographical variation in its ecological requirements, replacement in the
effect one factor has by another, or additional environmental variables in
other parts of the range. Thus, *Tilia cordata* is limited in the north by seed
sterility, in Britain due to the oceanic climate and in Finland to the
continental climate (Pigott, 1981).

In fact, the technique discussed describes range margins in terms of a
weighted sum of several variables varying independently in space, replac-

Figure 44. A similar geographical translation of the straight line separating sites inside and outside the range of *Carex pseudocyperus* in Fennoscandia. The dashed range margin is a reconstruction for the same species when the straight line is shifted +2°C; open circles indicate locations with subfossil remains during the warmer Hypsithermal (after Hintikka, 1963).

ing or compensating each other at one or more scales, or both. Ideally, it should be applied to observations throughout a species range (e.g. Caughley *et al.*, 1987). But ratios or product-moment correlation coefficients assume a linear response of species to environmental variables. Consequently, such results are valid for a small part of a species range only and cannot explain closed range margins. When closed, the intensities of one factor or factor combination compensate one another. Taking a complete species range into account rather than a part thus improves the analysis. Another improvement is to differentiate areas with diverging intensities within the range. For example, Carne (1965) plotted the

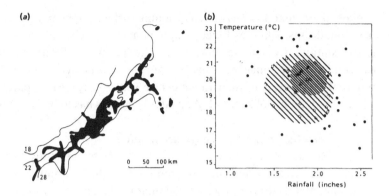

Figure 45. Range of *Perga affinis* in Australia, together with isopleths of some selected temperatures (*a*). Plot of locations inside (○) and outside (●) its range in a temperature–rainfall coordinate system (*b*). Hatched: expected range; heavy hatching: part of the range with highest densities (after Carne, 1965).

abundances of the Australian sawfly *Perga affinis* (Figure 45) against local temperature and precipitation for the period 1958–63. Two ellipsoids emerged, comprising sampling locations of two intensity levels, one containing high intensities lying within that of low ones. Sampling locations where the species is absent lie around these two ellipsoids. Using only the extreme intensity value of their range margin is thus an arbitrary choice; using other values reached inside the range gives a more complete picture (cf. also Rogers, 1979; Rogers and Randolph, 1986). As a further improvement, the intensity of pollen occurrences can be described in terms of more complex, algebraic polynomials comprising several environmental variables to which the species responds non-linearly (Bartlein, Prentice and Webb, 1986). Differences in range location in terms of climatic variation can be estimated using discriminant analysis (Caughley *et al.*, 1987).

Using algebraic equations does not add new ideas or information. Quantification makes our assumptions explicit, and allows us to change assumptions, to see the consequences, and to make generalization more obvious. For example, by attempting to formulate the species' intensity at a part of its range margin algebraically, makes us wonder what the exact formulation is and if this formulation holds for locations farther inside the range. If so, one has to assume a non-linear function to allow for the intensities becoming lower towards the margins, or to even explain these margins anyway. Optimum surfaces appear a truism to mathematicians, whereas biogeographers and ecologists often think of spatial stationarity. On the contrary, research either on limiting ecological conditions at range margins, or on population dynamics of local outbreaks of species in their range centre solves only certain parts of a species' probability of survival. At a later stage this has to be integrated.

Direct approaches to range delimitation

Results obtained by indirect approaches to the study of range margins are arbitrary and, at best, render hypotheses about particular environmental factors or processes. Similar to classifications and ordinations, they narrow down the range of possible explanations. As hypotheses, they must be tested. Two other drawbacks are that (1) many environmental variables are correlated with each other, and (2) organisms respond physiologically to these individual variables and to the effects of their interactions. Direct approaches through field observations or laboratory experimentation solve these kind of problems. For example, *Ilex aquifolium* becomes increasingly restricted to woodlands towards its range margin, which suggests that in the absence of tree canopy photosynthesis is critical to the species' survival in mild winters. This can be tested experimentally. *Ilex* may also be susceptible to frost, its cambium cells being injured by low temperatures, as happened in Denmark during the severe 1939–42 winters. This affected the northernmost populations most heavily or even killed them (Iversen, 1944). Field observations made in the 1963 winter in England showed that previous conditions may also be important, the western populations being more badly affected than the eastern ones near Cambridge where they could stand frosts as severe as −17°C during several nights (Pigott, 1975).

Common to these observations is the idea that extreme conditions may be important in limiting a species' range. The extremity of these conditions could be estimated by the deviation from their mean value. Yet, despite the fact that plants often remain unaffected by large deviations, differences between average conditions across boundaries often involve only 1°C or 2°C or less.

Among studies by Pigott and his associates, the study on the control of the northern limit of *Cirsium acaule* illustrates the complexity of range limitation best. Pigott (1974) concentrated on reproductive capacity at or near the range margin in terms of fruit production and seedling establishment. Conditions worsen towards the margin in both respects. After vernalization and a daily light exposure exceeding 15 hours, flower-heads of *C. acaule* begin to develop in June after which the rate of development increases with temperature. The first flowers open in late June in southern, eastern, and central England, but do not open until early August further north in Derbyshire. The fruits become detached five weeks after the outer flowers in the capitulum open, in which three categories can be distinguished: fruits without an embryo, with shrivelled embryos, and with fully grown embryos. Those without an embryo do not correlate with weather conditions, and those with shrivelled embryos correlate indirectly through the proportion of insect-pollinated flowers. Pollination chance depends on weather insofar as low temperatures restrict insect activity. Pigott (1974) does not mention if pollen germination and pollen-tube growth also depend on temperature, as in *Tilia cordata* (Pigott and Huntley, 1981). In *C. acaule* fruits with fully-grown embryos, growth-rate of the embryo increases directly with temperature, ripe fruit mainly occurring on the south side of the involucres, even in favourable years (Pigott, 1968). In dry conditions only its terminal capitulum develops, under moist conditions it produces three to five capitula, while it does not flower in the shade.

Cloudiness and precipitation affect reproduction indirectly through the temperature in the flower-head and enabling the fungus *Botrytis* to infect the fruits. The effect of precipitation on flower-head temperature was tested experimentally by spraying them with water, after which the ratio was estimated between numbers of calories required for drying the head and incoming energy during the day. *Botrytis* infects, particularly when cool and wet in late summer, and when *C. acaule* is in long grass. Wet weather also affects this species directly by preventing the capitula opening.

These factors affect the number of seeds, which becomes critical for colonizing new, open sites *C. acaule* favours. Thus, species maintenance through a continuous spatial dynamics requires a minimum seed-number. Field observations and experiments on fruit production should thus be complemented by studying the spatial population structure of *C. acaule* and the dynamics of this structure in relation to the spatial dynamics of its habitat, more dynamic environments requiring a higher seed production than more stable ones. As the flowering period in the south is longer and growth rate higher than in the north, yearly seed production is higher in the south than in the north. The effect of *Botrytis* infection is superimposed onto this, in many years resulting in a net production of 2–300 fruits in southern England and only 20–50 in warm and dry years in Derbyshire (Pigott, 1974). Species' survival is more at risk close to its margin than inside its range. It also reproduces vegetatively from rhizomes which may be up to 15–20 years old, and clones may even become 60–80 years. Moreover, English populations contain hermaphroditic plants besides female ones (Pigott, 1968).

When transplanted, *C. acaule* can persist for long periods, even on north-facing slopes where they have persisted for 12 years and spread vegetatively. Moreover, at northern sites not only is reproduction more difficult, but also there are fewer suitable, open places with a sufficiently high pH. On south-facing slopes fruits are produced and seedlings enable spread. Also, at the fringes of their ranges species may occupy, as a rule, more open habitats (e.g. Pigott and Walters, 1954; Carter and Prince, 1985), although, to my knowledge, this has not yet been tested statistically for a large flora. Species may live in more open habitats or as weeds towards their range margin because their vigour decreases in this direction, making the species less and less able to withstand competition from other species.

The weevil *Hypea postica* in North America is an example of a species that is limited by different factors in the northern parts of its range than in the south. Precipitation, in particular, enables fungi to affect the larvae, thus limiting the weevil's range (Cook, 1925). Michelbacher and Leighly (1940), working at its southern margin, concluded that high temperatures cause the species to aestivate, discourage the reproductive organs to develop, or even inhibit reproduction altogether. To the west, in California, conditions allow *H. postica* to occur at high densities all the year round, but a parasite, *Bathyplectes curculiones*, occurring there under optimal conditions, seems to keep densities low. This shows that (1) different environmental factors limit this range in the north, south, and west, and (2) its phenology and density under optimal conditions may differ from those near its range margin. Variation on several scales may also complicate matters. For example, the probability of *H. postica* larvae being attacked by a fungus increases northward depending on low temperature and high precipitation. The probability of fungal attack may superimpose part of a general trend of cooler and moister conditions towards higher latitudes.

This section shows that an intricate complex of several processes can limit species locally. Yet, we still have to try and assess the risk of extinction that marginal populations run. But despite their complexity, processes at range margins may be simple relative to those occurring in or near the range centre. The problem of why to study range margins is discussed in the next-but-one section.

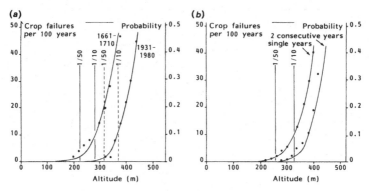

Figure 46. Risk distribution of oats as the number of crop failures per century as a function of altitude on a Scottish mountain. Drawn lines: expected risk distribution, dots observations. Figure 46 *a* gives the relationship for two climatically different periods and (*b*) that for one year and for two successive years during the whole period (after Parry and Carter, 1985).

Risk assessment

We need to assess the risk of dying out that marginal populations in particular run when the intensity of occurrence decreases from the range centre towards the margins and when the variance of population fluctuations increases in that direction. Parry and Carter (1985) estimated the yearly number of growing-degree days (GDD) for oats in northern Europe from the excess of the mean monthly temperature above 4·4°C. This base temperature was chosen because phenological studies on oats in Northern Europe indicated that below 4·4°C the plants show no significant responses. The time-series thus obtained covers 323 years over the period 1659–1981 (Figure 46). Several frequency distributions can be constructed from this time-series, expressing the frequency of the number of years with a low GDD-value to the number of years with high GDD-values, giving a low or a high risk of failure, respectively. The minimum amount of energy required was assumed to be 970 GDD, that is 90% of the mean accumulated warmth during one year. By considering temperature only, effects of precipitation on plant growth are not included, but this simplification can easily be rectified. For example, for barley on Iceland, Bergthorsson (1985) raised the required temperature sum by 30 for every 100 mm precipitation exceeding 200 mm during the vegetative period. We can thus estimate for each year whether oats will fail to give a yield at various parts of its range at or near the margin.

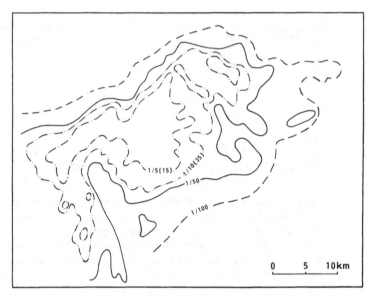

Figure 47. Probability of crop failure of oats for several altitudinal levels on a Scottish mountain (after Parry, 1978).

Where the frequency of crop failure is high, the risk of extinction is also high, and where it is low, the species will become firmly established. Parry and Carter (1985) calculated the risk of extinction and of establishment for various altitudinal levels on a Scottish mountain, assuming a temperature lapse rate of 0·68°C for every 100 m (Figure 47) (cf. Parry, 1976).

These calculations were made to estimate crop failures for oats at various altitudes, together with zones of risk intensity for oats of failing once in 50 years and once in 10 years. Low on the mountain, there is practically no risk in cultivating oats, whereas at high altitudes this risk is quite high and rises quickly. The same information can be mapped for various altitudes, the contour lines giving the estimated risk of crop failure of 1 in *n* years. As a next step, similar curves were calculated for the risk of crop failure in two successive years, being assessed from the same time-series of GDD. Obviously, two successive crop failures are rarer than failure in single years, but the effect may be worse. Just as farmers run out of money after several crop failures, individual plants or a species' seed or bud bank soon become exhausted.

From a species' physiology, we can predict its risk – or, more precisely, its survival probability – of occurring in a certain area. This procedure is more reliable than measurements for individual years, particularly if the plant has rhizomes, bulbs, buds, tillers, a seed bank, or suckers. All these morphological properties are ways of bridging unfavourable periods of a certain length, guaranteeing species' local survival. In principle, the same can be done for perennials and woody species by knowing their physiological resistance to frost, desiccation, etc. We can also build in the maximum yearly period of drought, possibly with its seasonal occurrence, to explain the local survival of perennials or trees. For certain areas this maximum may explain why they contain a high proportion of annuals. Thus, Van Zeist (1969) reasoned that in the Middle East farming was feasible as soon as precipitation during the autumn, winter, and spring allowed cereals and legumes to grow there. As annuals they could complete their life cycle before the hot summer, whereas trees could not do this there, despite favourable temperatures at the time.

No allowance has so far been made for dispersal, as seeds of crops are dispersed by man, or for the availability and dispersion of suitable habitats and their dynamics. But using it introduces dynamic processes into our considerations, which I leave until Chapter 10. Our considerations on range margins will be concluded by asking why relatively simple processes operative over small areas receive so much of our attention relative to that of the large area of the range as a whole.

Why study range margins?

There are three possible reasons explaining why the study of range margins has attracted so much attention. (1) They show how far the species' ecological requirements are in equilibrium with properties of the environment. (2) Although processes at the margins may prove to be complex, they are simpler than those nearer to the range centre. Biogeographers automatically select range margins as their study object because they are the simplest and progressively approach more intricate and complex processes. (3) Conditions at the range margins discriminate best between those favourable for the species and those that are not. Yet, I feel that a species' ecological requirements on a geographical scale do not show up sufficiently by studying its ecology at the range margins only, that is under extreme and possibly unrepresentative conditions.

Concerning the complexity of processes at the margins relative to those nearer the centre ((2) above), problems may be similar; some processes may be complex, and others may be simple. For example, Prince and Carter (1985) found that *Lactuca serriola* shows a graded response to summer temperature from south to north in England up to 22·2°C and 14·6°C during the day and the night, respectively – i.e. at Mediterranean temperatures – above which it does not respond to temperature. Above these temperatures, other factors become limiting, replacing temperature and explaining further intensity increases towards the centre, as with other plants and animals. Also, with an increase in the number of populations towards the range centre, the risk of inadequate dispersal capacity gradually declines, eventually hardly playing any role at all. As mentioned, observations on crops show that the variance of fluctuations decreases towards the centre, which reduces the population's risk of extinction. However, the probability of density-dependent processes keeping higher abundance levels within certain bounds may increase. Processes operative at the margins may thus differ quantitatively or qualitatively from those operative in the range centre (cf. Pratt, 1943). We cannot extrapolate or generalize our results from studies on one population to other ones somewhere else or to the species as a whole. Moreover, some factors or processes may affect all populations more or less uniformly over the range which can, consequently, not be discovered by studying populations or species. These are discovered by analysing patterns and processes described in Parts I and II.

These arguments particularly apply for populations that are both quantitatively and qualitatively in equilibrium with environmental conditions ((1) above). The sharpness of ecological adaptation at range margins illustrated above indicate that species are fairly well in equilibrium with environmental conditions. When they seem not to be in equilibrium, variation on different scales may complicate the picture. Before discussing matters of scale from this viewpoint, and sampling such an inhomogeneous and complex entity as a distribution range, I turn to another complex of factors, viz. those collectively known as biotic factors.

The geography of species interactions

The previous sections considered the effects that abiotic factors can have on the structure, shape, size, and delimitation of species ranges. How can we do justice to biotic effects in this way? Two kinds of effect can be distinguished: (1) a species may be absent because another species on which it depends is absent, and (2) the presence of one or more other species affects the intensity level of the species concerned.

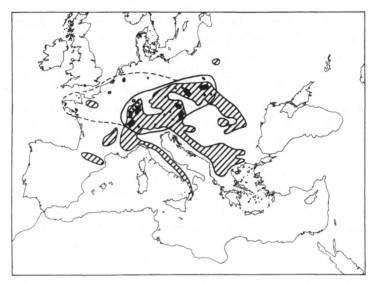

Figure 48. Realized range (heavy line) and potential range of *Choristoneura murinana*, a monophagous insect on *Abies alba* (hatched) in Europe (after Eidmann, 1949).

Potential ranges through monophagy

For eco-physiological reasons one sometimes expects a species to occur abundantly in one region, whereas in fact it can hardly be found there, if at all. For example, a monophagous parasite may be absent in areas where its host is absent. *Choristoneura fuminifera*, which is a monophagous insect on *Abies alba*, only occurs in the western part of its host range in Europe, whereas outside this range it is absent (Figure 48) (Eidmann, 1949). Unlike other parasites it does not shift to other potential hosts. But as soon as *Abies alba* was planted outside its natural range to the west, *Choristoneura* extended its range to these areas where it may presently reach outbreak densities. The region of the insect's highest outbreak frequency thus coincides only with the range margin of its host. In the host's range centre the probability of *Choristoneura* developing to outbreak proportions is low, indicating that this area is ecologically marginal for it. This suggests that the range of *Choristoneura* coincides only partly with that of *Abies alba*, although outside the range of the tree, for other reasons, the parasite potentially has its range centre. One may call the realized part of its range the species' effective range and the area where it could occur, but where it in fact is lacking, its potential range (Franz, 1964).

From a biogeographical viewpoint the possibility of distinguishing effective and potential ranges is important, here it is interesting in connection with the range size of a host relative to that of its consumer species, that of the host generally being the largest (e.g. Strong, Lawton and Southwood, 1984). Apart from biotic factors, the locality and dynamics of the ranges are principally determined independently of each other by climatic factors. Consequently, in different parts of its range a host species can be affected by different combinations of consumers.

Changes in the general level of intensity

Locally, species density may be altered by a biotic factor such as disease, parasitism, predation, or competition. Without this densities would be high, whereas in its presence the species may even occur endemically. For example, local densities of *Hypericum perforatum* in California are kept low mainly by the beetle *Chrysolina quadrigemina* (Huffaker and Kennett, 1959). Both host and parasite are native to the Old World, and were introduced into North America at different times. *Hypericum* was introduced at the turn of the present century and *Chrysolina* in 1945–6, together with its less successful congeneric species *C. hyperici*. After its introduction, *H. perforatum* became extremely abundant in open, sunny and well-drained areas, even eliminating the native flora. The beetles were introduced to control *Hypericum*; their larvae defoliate the plants in winter and spring, thus preventing them building up reserves for the dry summer. Their introduction was so successful that at present *H. perforatum* occupies shady areas only, whereas the beetles prefer to lay their eggs in sunny sites where the plant was initially a pest. This also happened in Australia where *H. perforatum* now also occurs in forests despite its preference for open habitats where the beetle *C. gemelata* feeds on it, leaving it untouched in forests which the beetle is unable to inhabit (Clarke, 1953). Also in Australia, the introduced cactus *Opuntia* spp. initially ruined millions of acres of pasture and rangeland. It occurs only temporarily in scattered populations after the subsequent introduction of the moth *Cactoblastis cactorum* (Dodd, 1959).

This phenomenon can also be observed on a broader scale. Similar to the almost total disappearance of the American chestnut, *Castanea dentata*, caused by chestnut blight during the first half of the present century (Anderson, 1974), hemlock (*Tsuga canadensis*) suddenly collapsed about 4800 years ago all over its range in north-east America (Davis, 1981*b*). We cannot predict the fate of the chestnut on the long term, viz. whether it can eventually recover from this disease or not. Similarly, we cannot tell what will happen to elm in Europe over vast areas having been exterminated by Dutch elm disease. We only know that it took hemlock about 2000 years to recover and reach stands with a lower density than it had before its decline. Not only is it impossible to explain present patterns of distribution and abundance exclusively in terms of climatic conditions, but also it is impossible for the period prior to the decline of hemlock for patterns in birch and oak. Rapidly increasing amounts of pollen show that immediately after the hemlock decline, birch may have taken the vacated places, later followed by a similar increase in oak. This also happened recently after the chestnut decline (Brugam, 1978; McCormick and Platt, 1980). These tree species can potentially reach levels of intensity different from the ones observed, their observed levels being determined by abiotic and biotic factors jointly.

Temporal factors complicate the picture. For example, Messenger (1970) studied the phenology of hosts relative to that of their parasites. Regarding the impact of biological control for forestry, it is often thought that species' phenologies should be synchronized to give optimal success, a prerequisite that is not always met, and if it is, it does so in part of the host's range only. This agrees with Strong *et al.*'s (1984) statement that more than one biotic agent can affect a host; they are all affected individually by different climatic factors.

These examples illustrate how local habitat preference can be determined from field observations. Abiotic conditions allow *H. perforatum* to grow preferentially on sunny places, but biotic factors completely prevent it from doing so. In turn, the beetles soon only find scattered plants to eat. This results in them occurring at lower densities than those resulting from abiotic conditions alone. Moreover, in British Columbia *Hypericum* fluctuates irregularly and at higher levels than in California because *Chrysolina* is not adapted to more northern climates (Harris, Peschken and Milroy, 1969). Observed distribution and abundance patterns may thus give a wrong picture of those that could be reached without biotic factors operating. It is hazardous to rely on correlations between a species' local and geographical occurrence and abiotic variables only, although these may be of primary importance on a broader scale. Depending on their number, relative weight, and independence, effects of individual factors – effects of a herbivore or competitor, for example – may look more profound, but usually this happens on a relatively fine spatial scale only. On a fine scale the effect of a herbivore or competitor may even be spectacular, as in *H. perforatum* being defoliated by the beetle species, or in possible competition between the grey and red squirrel in England (Reynolds, 1985). But on a broader scale they appear to be of restricted value. And on a geographical scale the distribution of species ranges may appear independent of each other (e.g. Pielou, 1977). There is nothing contradictory in finding ranges being explained by abiotic density-independent factors and local populations by density-dependent biotic ones (Enright's (1976) biogeographic dilemma) as long as we take account of differences in the scale of variation of both species and their environment.

Water-tight distinctions between effects of biotic or abiotic factors are impossible to make at geographical scales because of diffuse competition (MacArthur, 1972), or, including all biotic agents, diffuse interference. This seems realistic, although testing hypotheses on the importance of biotic factors in determining the size of populations or species ranges is usually practically impossible. Diffuse interference would lower the intensity level of optimum surfaces over the whole range of distribution and could conceivably result in smaller range sizes. If the numbers of biotic factors are large enough, of equal weight, and vary independently, they exert a uniform and constant pressure all over the range. The ranges maintain their individuality of responses to climatic variation. None of these assumptions hold in all cases and one cannot suppose that they apply in conjunction. That many species interactions may result in a general lowering of intensities over the range may be most apparent when their effects are excluded as may happen in species introduced into alien regions or continents (Dobson and May, 1986). A picture of a general lowering of the species' intensities over its range, whereby the range structure depends on responses to abiotic factors, makes the tapering from high, central intensities towards low ones at the margins understandable. It is difficult, if not impossible, to visualize the same pattern resulting from effects that one or more species exert on each other, whereby these species even overlap only partly and show independent range dynamics over time.

Explaining why species ranges have a particular size, structure, and location cannot solely be done in terms of abiotic factors; we must also know the species' physiology, anatomy, morphology, behaviour, and demography, as well as its possible interferences with other species. It is, in principle, the complexity of all these properties that determines the size, structure, and location of species ranges, as well as the temporal variation in these statistics. Alterations in one or more elements of this complex will affect one or more statistics characterizing the distribution range. Conversely, similarity in this complex among several species can be reflected in a similarity of these statistics.

Invading species

Although invasions are an aspect of species dynamics to be treated in the next chapter, they contain some features that are relevant here. The behaviour of invading species is often examined to estimate the relative importance of abiotic versus biotic factors. Such species are supposed to lack natural enemies to control their numbers. First, a species would expand in all directions at more or less the same rate, a process that can be described in terms of diffusion, together with either exponential or logistic growth of their established populations (cf. Hengeveld, 1989a). Skellam (1951) effectively applied this model to the spread of muskrat in Europe and of the oak in Britain after the retreat of the Pleistocene ice.

Applying two-dimensional diffusionary models, one assumes no barriers or factors operative that delay the process, that is that interference from species in established communities plays a minor role, if at all. The idea is that invaders meet a growing resistance only later, the established species gradually adapting to them (e.g. Elton, 1958; McKillup, Allen and Skewes, 1988). Yet, it is not certain how many species arrive in another area, on another continent or island, how many fail to become established, and why they fail. Reviewing the literature, Simberloff (1981) was unable to show species replacement or limiting similarity among species, suggesting that, possibly except in a few instances, interspecific competition does not represent a major force in structuring communities. Similar conclusions were drawn by Crawley (1986) and Lindroth (1957), both working on extensive, presumed unbiased data sets on established species. Species interference thus seems not to hinder species invading a new area. Another hindrance may be that, for a species to establish itself, its populations must reach a minimum density. But these authors could not find evidence for this effect either. Similar conclusions can be drawn for species deliberately introduced to control pests (Crawley, 1986). Simberloff (1981) suggests that only general guide-lines can be given for the success of introducing species, such as a similarity of the climatic conditions in the new home country to that of the native one. Climatic conditions may be of prime importance to the probability of establishment and eventual success of invading species. Climatic conditions alien to the species may also explain why most (e.g.

Lindroth, 1957; Mayr, 1965*b*) invasions fail. Conversely, it is not known why it took the starling, after several unsuccessful attempts to introduce it from Europe into North America and twenty years without a significant increase, to build up its numbers in the next forty years to over 50 million individuals.

This raises two questions: (1) can species be predicted to be invaders from known properties, and (2) what kind of communities do they invade? From a statistical analysis, Crawley (1986) suggests that the best invaders are those with high rates of increase, and are widespread and numerically abundant rather than patchily distributed and rare in their native areas. The environments in which most invaders occur in Great Britain are those with a high proportion of bare ground or are frequently disturbed, although no community is closed completely. They can all be invaded.

Despite these general guide lines, prediction of a species' success as an invader or as an introduced species into certain areas is still impossible; too much is unknown (cf. Simberloff, 1987). Also, apart from the characteristics of the species and the communities and habitats them-selves, prediction depends on stochastic factors, such as numbers of individuals involved, number of invasions, the heterogeneity of their environment and gene pool, and on the timing of their arrival relative to environmental conditions. Effects of competition, though important in individual cases, cannot be considered to represent factors generally affecting a species' probability of invading a particular area. The balance still seems to indicate that abiotic factors outweigh biotic ones.

Optimum-response surfaces and climatic reconstruction

Recent efforts to reconstruct past climates have given insights into the nature of species ranges in terms of optimum-response surfaces (e.g. Bartlein *et al.*, 1986). But in these studies knowledge from bio-geographical patterns is, in fact, used to reconstruct climate (cf. CLIMAP, 1976; COHMAP, 1988; Huntley and Prentice, 1988) rather than that climate predicts these patterns. Essentially, they belong to climatology rather than biogeography; species are used as climate indicators.

Since variables operating on different spatio-temporal scales will affect species differently, the reverse is also true: different climatic variables can be expected to be detected using different scales of biogeographical sampling. Because of their greater mobility and shorter life-cycles beetles, for example, give a much finer resolution than trees (e.g. Coope,

1975). Similarly, temporal resolution in ocean temperatures during the Quaternary is limited because benthic organisms continuously mix the upper stratum of the ocean bottom. Tree rings, in their turn, are more suitable for reconstruction of fine-scale variations, as they record local conditions on a year-to-year basis. On the other hand, certain climatic processes are steered by long-term variations in the earth's rotation and its movement around the sun, but others can depend on the temporary location of warm or cold patches in one or more oceans (e.g. Lough, 1980). Other effects depend on interactions between climatic variables and local soil variation or topography and yet others are best described in stochastic terms (e.g. Stern, 1980). Therefore, depending on the scale and climatic process of interest, one must choose a particular taxon. But the inferences one draws pertain, as a consequence, to that scale of variation only, rather than to finer or coarser ones.

A similar reasoning applies to climatic resolution in space. Small ranges can only indicate spatially fine-scale variation and large ranges coarse-scale variation. Special problems arise for elongated ranges. Often the principal geographical axes of elongated ranges do not have the same direction and extent in other species, thus suggesting different environmental variation. This problem is implicit in sampling stratifications of species ranges. For example, when two spatial gradients exist, one from moist to dry and one from cool to warm, it depends on the direction of the main trend of the range or on that of a sampling transect within a range relative to these two gradients what variable(s) can be detected. Using a species as an indicator therefore requires information about the topography of its total geographical range or of the part sampled within it.

Finally, although we may infer a particular environmental variable to account for the location, size and shape of a species at a certain time, without experimental evidence we cannot be completely certain that our choice from all possible variables is appropriate. In fact, Imbrie and Kipp (1971) did no experiments on temperature preference of foraminifers after inferring from factor analysis that temperature could be an important determinant for their geographical distribution. Thus, their inferred temperature values may, in principle, just substitute the geographical coordinates of their sampling locations. Moreover, experimental evidence may allow for non-linear responses of a species to particular variables, thus enabling more climatic information to be extracted from a species range when spatial variation in abundance is known.

Knowledge of species ranges can help to reconstruct climate, especially when ranges are interpreted as optimum-response surfaces, but lack of knowledge of their structure, dynamics, and causation should discourage rash inferences of climatic variation, or of species that respond to this variation.

Conclusions

It is necessary to sample species at several spatial and temporal scales simultaneously, and to characterize the compound fluctuation pattern by risk functions for particular periods. These risk functions, from the species' viewpoint, depend on properties such as longevity, dispersal capacity, germination time, drought resistence, and seasonal temperature or precipitation requirements, for example. Our objective in biogeography, therefore, is to determine how far spatio-temporal variation in a species' life-cycle, dispersal capacity, physiological requirements and resistance, and population dynamic parameters match environmental variation in space and time. Problems arising for a species from mismatching environmental variation may decrease towards the range centre, as more and more factors cease to be limiting. This may explain its increasing intensities towards the range centres, sometimes causing a species to become a vigorous pest.

All things taken together, species ranges can be considered as optimum-response surfaces with a complex internal structure. It is not possible to obtain insight into this structure from samples at a few locations only, nor from a great number of them taken in only part of the range, or at one scale. By discussing range structure, we have only paid attention to spatial variation. The next chapter will add the dynamic component to this picture.

10

The dynamic structure of species ranges

The picture of high intensity values in the range centre and low values towards the margins can represent a spatial frequency distribution of values that various local population dynamic parameters reach during a certain period; within this period local values can fluctuate widely. In this chapter I discuss the way that spatial frequency distributions build up over time.

Good's Theories of Tolerance and Migration extended

Good's Theory of Tolerance (1931) and Theory of Migration (1974) together consider the same phenomena described here as the structure and dynamics of species ranges. Yet, they are not identical but can be extended. In the Theory of Tolerance a species' geographical range is identical to that of the range of its ecological tolerance – or ecological amplitude – and thus to the statistical concept of range. But, statistically, a range is one out of several measures for characterizing variation. Ecologically, it is also insufficient to characterize species by living conditions at their range margins only. Instead, we have to describe species ranges in terms of internal, structuring processes of which we can distinguish (1) physiological responses to environmental gradients, (2) changes in these gradients, and (3) changes in intensity distribution following those environmental changes.

Non-linear responses to environmental conditions are best known from gradient analysis (Whittaker, 1967). Here, a species' response would follow an optimum curve, in which each point represents the asymptotical value K of the logistic growth equation (Gause, 1932). Thus, the various intensity levels can be thought to be physiologically controlled. However, as each point represents a local intensity value, the process is spatially static; it does not describe movements of individuals or propagules from one location to another. To extend Good's Theory of

Tolerance further, I assume that (1) the environmental conditions along a gradient or over a two- or three-dimensional space change over time, and (2) the intensity distribution of the species follows this environmental change.

At each location the values of several independent and unweighted variables can be thought to vary randomly over time, to a certain species some being favourable and others unfavourable. When certain favourable variables have a particular probability of coincidence, they will form favourable spatial clusters or nodi, which will also happen with unfavourable ones. Species, in turn, often respond individually and non-linearly to these nodi of environmental favourability. This pattern depends on the number and variability of the variables concerned, on differences in their relative importance, and on mutual dependences of species. Depending on geographical gradients in the nodus frequency, as well as the species' response, the overall favourability of a species' environment will generally increase towards its range centre and decrease towards its margin.

Species must match this spatial system of stochastically shifting, ecologically favourable and unfavourable nodi. This matching process may follow another stochastic process, which assumes mobile individuals or propagules and a dependence of the probability of the time at which it leaves the locality. When a locality is favourable, it assembles more individuals than unfavourable locations will, on average, do. The spatial pattern of transition probabilities thus defines the spatial pattern of expected intensities of the species concerned. Of course, mobility is relative, depending on the degree of the species' sensitivity to changes in its environment, as well as on the scale of investigation. Moreover, mobility determines the probability distribution characterizing the chance that an individual reaches a locality at a certain distance.

When environmental favourability of various locations is similar, and when the transition probabilities are normally distributed, the spatial movements of propagules or individuals are Brownian and the resulting spatial process becomes diffusionary, describing flows of individuals from localities of high density to those with low density (e.g. Skellam, 1951). Particularly in the latter case, individuals would continually trespass the range margin, often succumbing there, but occasionally establishing, thereby extending the range in that direction. Conversely, organisms may leave or die in certain parts of a range, thus making the range contract locally. As this is basic to Good's Theory of Migration for cases where extensions and contractions occur at more or less opposite ends of a range, range shifts are a direct consequence of its internal dynamics. Moreover, when the contraction rate at one end equals that of extension, range size stays constant; when they differ, the range either enlarges or becomes smaller. Thus, range extensions and invasions are special cases of a more general, continuous process of spatial adaptation to ever-changing environmental conditions.

For mobile species it is possible to explain the location, structure, and dynamics in stochastic terms. But spatial adaptation is just one of the possibilities for coping effectively with the vicissitudes of a changing environment. There are other possibilities, allowing species to stay where they are for a longer time. Some present-day patterns can be explained by patterns and conditions of former times, that is, by historical processes. Yet, when all mechanisms fail, species find themselves in conditions adverse to them, causing them to die out.

In this way, Good's Theory of Tolerance and Theory of Migration can be extended and integrated from a description of a species' geographical range in terms of physiological tolerances to that of the geographical frequency distribution of intensities in terms of both physiological optimum surfaces and spatial adaptations. In this chapter we shall first look at the mechanism of climatic variation and, after this, at the way species match this variation.

Climatic release and dispersal in the spruce budworm

A species' spatio-temporal variation matches climatic conditions by means of specific morphological, behavioural, and physiological adaptations. Its continuous spatial adaptation to a geographically highly dynamic environment, so essential for its survival, cannot adequately be described from average locations or densities, or from yearly range limits. Neither a species' properties and their evolution, nor its survival and extinction probabilities can be understood when species and climate are looked upon as static patterns, rather than as dynamic processes. Their patterns are never the same, and if they are, they should be described in terms of dynamic equilibria of essentially stochastic processes.

For example, two eastern Canadian insect species causing damage to timber are the forest tent caterpillar, *Malacosoma disstria*, and the spruce budworm, *Choristoneura fumiferana*, both of which are known to have long time-series. Much local information exists about outbreak epicentres from their early developmental stages onwards. In the local time-series, population build up usually follows several consecutive years with a certain weather type (Wellington *et al.*, 1950); low temperatures and

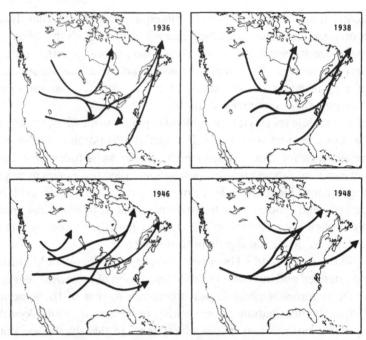

Figure 49. Courses of storm tracks across North America for various years (after Wellington, 1952).

high humidity favour the forest tent caterpillar, whereas dryness favours the spruce budworm (cf. also Greenbank, 1956, 1957; Greenbank, Schaefer and Rainey, 1980). Climatically, years with a prevailing weather type will have many cyclones passing over the insect's epicentre, which can be classified into three groups according to their region of origin, with cyclones from (1) arctic regions, (2) desert regions, and (3) maritime regions (Wellington, 1952, 1954). Air from arctic regions is usually cool and dry; continental air from North American deserts is warm and dry; and maritime air from the Caribbean is usually warm and moist on its arrival in eastern Canada. Local conditions can thus be characterized by multimodal frequency distributions of, for example, temperature, each mode relating to a particular region of origin (Figure 57) (Bryson, 1966).

Cyclones from one of these regions can also be counted directly, giving differences between average paths followed in various, subsequent periods (Figure 49). This varies with time for the three regions. The chance of an outbreak developing in four years can be calculated from local outbreak frequencies, which, in turn, are determined by the number of cyclones from a certain area of origin. Just as the local frequency of intensities of climatic elements can be characterized statistically, so can the chance of an epidemic developing be described in statistical terms. The frequency of outbreaks occurring varies in space, and may, according to the Central Limit Theorem, be more or less dome-shaped, the highest frequencies occcurring in the centre and the lowest at the range margin. However, local abundances cannot be explained only by local and instantaneous conditions, but by the spatio-temporal fluctuation regime of a range of intensities, together with the pattern of local winds. Often the intensity regime of several environmental factors may even coincide. Climatic conditions matched by a species thus form a multivariate frequency distribution, though of several partially related factors, each having its own spatio-temporal pattern and scale of fluctuation.

Greenbank *et al.* (1980) (cf. also Wellington *et al.*, 1950) described the regional outbreak development of the spruce budworm and the role of dispersal in this process. He found that outbreaks develop in about 4 to 5 years of dry, sunny weather that polar air masses bring into eastern Canada. Because of these favourable conditions directly affecting the insects and through flower production by the host trees, average female fecundity increases. Abundant flower production in dry years leads to rapid larval development, which, in turn, gives rise to more fertile females. Conversely, under these conditions, young females are less fertile when they had fed as larvae on old foliage and they develop less rapidly. Thus, both direct climatic influences and indirect ones through its host, cause the spruce budworm to develop from an endemic state to epidemic proportions.

Local outbreaks can also be initiated by immigrants from surrounding areas. For example, the 1949 outbreak in New Brunswick followed one in 1945 in Quebec 250 miles away, which approached the new area stepwise by several infestations during the intervening years. Voluntary dispersal cannot account for this spread. Rather, spread is involuntary by convectional and turbulent air, which transfers large parts of moth populations, particularly unspent females, at great distances from one locality to another, where the moths are dropped in often extremely high concentrations (Greenbank *et al.*, 1980). When conditions in this newly infected locality are suitable, the moths form a nucleus of a new outbreak or enhance the build up of an already developing one (Clark, Jones and Holling, 1978).

Climatic causes of range dynamics

So far, I have mentioned average routes that storm tracks take, as well as temporal variation in those averages. This points to some ideas that are important in present-day climatology. Of these ideas, I discuss only several aspects relevant to the dynamism of biogeographical patterns.

Two principles emerge: (1) tropical regions warm up, whereas those at higher latitudes cool, and (2) the speed of daily convolution of locations at different latitudes differs. The differential warming up of locations at different latitudes is caused by a decreasing inclination of the sun's rays from 90° at or near the equator to 0° at the Poles, which results in a system of latitudinal convection belts. Where equatorial air, for example, meets that from subtropical regions at the Intertropical Discontinuity, it rises and diverges again into a northward and a southward branch in the upper strata. South of this Discontinuity, humid, southern air streams prevail in the lower strata and north of it slightly colder, dry air flows south. At higher latitudes cold, arctic air flows south, at some point meeting a warmer northward flow, and together these rise to higher strata. At both latitudes the rising air gradually cools off, resulting in cloud formation and subsequent rain. Thus, rain falls particularly south of the Intertropical Discontinuity, hence the location of the Sahara desert and the tropical rain forests north and south of it, respectively. The convection belts at higher latitudes are less sharply delimited and more variable than the intertropical one and can be recognized less easily.

When air flows from one latitude to another, it retains its former momentum, which is high at equatorial latitudes and gets lower towards the poles. This generates westerly and easterly directions in the convectional currents at different latitudes. Thus, broad-scale cells of horizontal air flows develop, apart from the vertical currents of convectional air, resulting in a global system of air movement in dynamic equilibrium. This equilibrium, and thus the topography of the geographical pattern of air currents, depends firstly on heating by the sun, and secondly, on the momentum of the air. Thirdly, it depends on geomorphological features of the earth, such as relative location and reflective abilities of continents and oceans and the location, height, and direction of mountain chains that obstruct the flow of air.

One feature resulting from the latitudinal air flows having longitudinal momentum is the mid-latitudinal circumpolar vortex. This vortex is the statistical trajectory followed by the westerlies, or, ultimately, the jet streams, which are strong winds in the upper air around the pole. This flow varies between geological and historical periods, between and within seasons, as well as daily. It does not follow a particular latitude, but depending on, among other things, geomorphological features, it meanders with varying amplitude and location of the various constituent loops according to wind velocity. The location of one of these loops seems to be relatively fixed; when the air reaches North America from the Pacific, it is obstructed by the Rocky Mountains, causing it to flow north, after which it dips south from Alaska into eastern Canada. As there is no other major obstruction in its way, the location of the meanders east of the Canadian trough in Europe and Siberia, as well as their seasonal variation, can be described relative to this Canadian one. In turn, the topography of the jet stream both determines and stabilizes the trajectory of cyclonic and anticyclonic disturbances in the lower atmospheric strata, variation in its topography accounting for local weather fluctuation. In the temporal pattern of these fluctuations, fine-scale variations, such as cyclones, are superimposed on broad-scale ones, the jet streams, which, in turn, vary on various large time scales. The location of the circumpolar vortex is at the equilibrium point of arctic air streaming south and temperate air streaming north, and is thus ultimately determined by the atmospheric energy balance. Because of latitudinally differentiated heating and cooling of the atmosphere, temperatures are unevenly distributed, forming latitudinal and altitudinal gradients, the steepness of which determines the latitude at which the circumpolar vortex will occur; the steeper the gradients, the lower this latitude, and hence the larger the cold arctic areas encircled by the vortices. In winter the gradient becomes steeper, causing the vortex to move south and to bring cold air to lower latitudes. Apart from seasonal variation caused by yearly variation in inclination of solar radiation, other factors also affect the steepness of the temperature gradients. Variation in their values also influence the latitude of the circumpolar vortex, resulting in larger parts of the earth being influenced by cold air, whereas smaller parts are warmed by tropical air when it occurs more to the south in summer.

Because of the continuity of these processes, variation in the location of the circumpolar vortex could also be continuous, but in fact it is not; it varies discontinuously. As mentioned, the Northern Hemispheric vortex meanders, one of the loops often being fixed by the Rocky Mountains. As the number of loops can only be an integer, the latitudinal course of the vortex varies stepwise, the greatest number of loops being allowed by the greatest vortex length. This discontinuity, both in time and space, has considerable biogeographical consequences, explaining, for example, the spatial discontinuity of floras and faunas (Chapter 12).

However, these air streams pertain to averages about which considerable spread occurs. For example, the course of the loops of the northern circumpolar vortex is unstable; their amplitude varies daily and they shift eastwards. Together with variation in amplitude, wind speed varies; lower speeds give wider loops. Only occasionally can a loop be so wide that its topography stabilizes, or 'blocks', making local weather constant for one or a few months. The proportions of air types within and over the years determine together with other ecological factors, the possibility of a local build-up of population density, or more generally, the location, size, shape, structure, and dynamics of species ranges. Thus, the build-up of an epidemic of the forest tent caterpillar in eastern Canada only occurs when weather is influenced for 4 or 5 years by warm, humid maritime air from the southeast. Conversely, epidemics of the spruce budworm can affect the same region when cold and dry polar air enters it, also for 4 or 5 years. Where and when outbreaks develop depends on the way the air masses keep in balance over the years, which also depends on other air masses. This means that daily weather varies locally, although on average either more polar air or more maritime air enters the region. The fact that local observations of both climate and species intensities can only be described statistically, reflects their dynamic nature.

Features of range dynamics

Species responses to climatic spatio-temporal variability by means of physiological and spatial adaptability may explain several phenomena. First, mobile species can exist under erratic environmental conditions by means of either a high reproduction rate or by spatial exchange of individuals. This results in rapid shifts in density distribution within the range and in range margins. In only a few years marginal populations with low, endemic density levels can develop to outbreak proportions. This happened, for example, in some Central European insect species during the climatically continental period of the 1930s and 1940s, after which they returned to their previous endemicity (Pschorn-Walcher, 1954; cf. also Hengeveld, 1985*b*; Kozar and David, 1986).

Continual shifts in range location can also explain the existence of disjunct relict populations of or around these ranges, or the often high level of endemism in mountain regions. Mountainous environments vary considerably within short distances, a comparable variation only being found over large distances in flat countryside. Adapting spatially to climatic variability, both horizontally and altitudinally, requires shorter distances to be covered in mountainous regions than in flat land, thus becoming feasible for species with restricted dispersal capacities. To such species belong, for example, many trees, such as sequoias or cycads, and mammals, with extensive ranges during the warmer Tertiary, but which gradually became smaller and more fragmented under the cooling climate (cf. Patterson, 1984; see also Davis *et al.*, 1988). Thus, apart from shifting latitudinally, where their smaller numbers make it impossible for them to cover large distances, they often withdrew to the mountains.

Finally, species must be so variable that they can match environmental variability. Apart from the mean intensity level of environmental factors, the range, variance, and other measures of their ecological amplitude is significant, euryoecious species having the highest survival probability. The mean intensity of a factor can relate positively to a species' spatial occurrence; the smaller the probability of occurrence, the more endemic and vulnerable the species.

Thus far, I have assumed that species are so mobile that they can follow climatic capriciousness in space (cf. Ford, 1982), although this certainly does not apply to all species. Many plant species, for example, have restricted means of dispersal, if at all, preventing them from adapting to climatic vicissitudes by dispersing to other localities. Rather than escaping the new, unfavourable conditions, they must be either physiologically flexible or able to bridge the unfavourable period as seeds, buds, or rhizomes. When conditions improve, they germinate, bud, and reproduce. Yet, the unfavourable periods should not last too long, and should occur at more or less regular intervals. Immobile species should thus match, in another way, the frequency and duration of unfavourable periods. Just as mobile species, they should match the statistics of frequency distributions that characterize the fluctuation regime of favourable and unfavourable periods. Over the years they will also show local frequency distributions of intensity of occurrence similar to the more mobile ones.

Seasonality and optimum-response surfaces

One aspect of a species' local intensity is its seasonality, this being the probability of occurrence of events recurring within a certain part of the calendar year. This relates temporal frequency distributions within a year to geographical distributions. One such event is the onset of flowering, budding, or shedding its leaves. In biogeography it is of interest whether or not the event varies in space and time, as this variation may add to the species' probability of survival. The relevance of

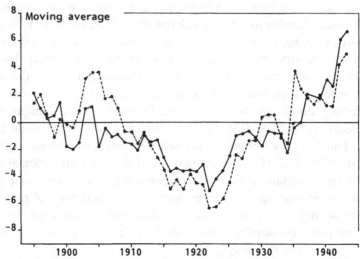

Figure 50. Moving average (days) of onset of flowering of hazel,
Corylus avellana (solid line), and coltsfoot, *Tussilago farfara*
(dashed line), in England (after data in Jeffree, 1969).

seasonality is that in order to survive, species should pass through several
life-cycle or developmental stages within a year; if they cannot, reproduction will fail, leading to local extermination, or total extinction. Species
respond regionally (e.g. Reader, 1983) and specifically to this variation;
during the relatively cold period 1910–30, the 10-year moving averages of
the onset of flowering of coltsfoot were lower than those of hazel,
whereas before and after that period they were higher (Figure 50)
(Jeffree, 1969).

The possible effect of latitude on seasonality can be estimated, for
example, from the average month that ground-beetle species were
recorded in the Netherlands, Denmark, and Scandinavia. The average
month of seasonal occurrence is significantly earlier in the Netherlands
(6·1 months from 1 January) than in Denmark (6·3) and Scandinavia
(6·3), whereas in the latter two regions it is the same. This tendency,
though stronger, also occurs in the length of their seasonal period, the
coefficients of variation being largest for the Scandinavian data (0·71) and
smallest for the Dutch data (0·32). This implies that, statistically, the
seasonal period increases with latitude.

Much less is known about the complete seasonal periodicity of species
than about bud or flower initiation, although these may be of smaller
interest. For example, Lindroth (1949) reasoned that the slightly higher

specific temperature of chalk allows some southern ground beetle species to complete their life cycle on the island of Gotland, whereas this is not possible on the granite on the Swedish mainland nearby where they are absent. A temperature rise of 2°C extends their activity period (or growing season in plants) by about 2–3 weeks, which would suffice to allow completion of the life cycle (see Sjögren, Elmberg and Berglind, 1988, for behavioural responses in relation to reproduction at their northern range margin). The length of the growing season also varies annually in Scotland, being relatively short up to the mid-1930s (Gloyne, 1972). This trend is also apparent in Finland, causing plant species to contract or extend their ranges extensively in a relatively short time (Erkamo, 1956). Similarly, in historical periods, the length of growing season allowed oat to grow at particular altitudes (Parry and Carter, 1985; cf. also Parry *et al.*, 1988).

Species responses are usually not simple and instantaneous, but complex. In long-lived species such as trees, individuals either bridge unfavourable periods, or live on a storage built up during a favourable period, taking several years to recover. Moreover, it often takes a population several years to build up from endemic to epidemic levels, or to disperse from one area to another. Thus, both individuals and populations carry their history with them and do not follow environmental changes closely. This is also true for responses to climatic variation within years, as individuals respond differently to the same environmental factor in various stages of their life cycle. Moreover, as species often respond to cumulative amounts of energy expressed as temperature sums or heat units, they do not respond to environmental fluctuations instantaneously. Also, the sequence of, for example, high and low temperatures over the season can be important, the wrong sequence restricting a species' performance. Thus, a warm spring followed by a cool summer gives a better crop of peas in Ethiopia than a cold spring followed by a warm season (Wang, 1960). It is therefore necessary to differentiate between growth processes in which organisms just get bigger and developmental processes in which the fate of future performances is determined (Figure 51) (Wang, 1960). For example, soil temperatures during the first two weeks of planting affect subterranean ears in corn, lower temperatures causing a longer period of maturation later on.

Organisms are energy-transforming systems, the properties of which differ between species and between life stages, which is, among other things, expressed in their local numerical abundance (e.g. Short *et al.*, 1983). Thus, spatial variation in daily and seasonal climate utilization

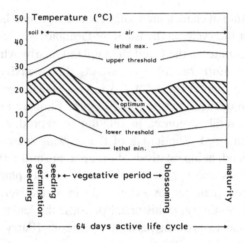

Figure 51. Temperature requirement of peas from seeding to maturity (after Wang, 1960).

determines a species' local and geographical dispersion. Since weather varies annually, a species' intensity and, hence, its probability of survival, will vary accordingly. One way to avoid the risk of extinction is to be physiologically flexible, another is to bridge unfavourable periods as seeds, buds, etc., and still another way is to be spatially adaptable. In all cases seasonality is essential for the extent that species match the variability of their environment and hence their survival probability reflected by their geographical distribution.

Extinction

A species' spatial response to its environment is usually imperfect, not only because of limitations in dispersal efficiency and population dynamic parameters constraining its dynamics, but also because species are physiologically flexible and genetically variable and adaptable. Yet, conditions can be too harsh and the time required for adapting to new conditions may prove too short, causing species to die out. The rate at which this happens will be difficult to predict, because of the number and complexity of the variables involved. Extinction is a stochastic process, because of the nature of environmental variability and the resulting dynamic structure of species ranges; numbers of individuals or populations are important determinants of the exact probability. The greater the range, dispersal efficiency, reproductive power, number of populations and ecological amplitude, the smaller the probability – and hence the rate – of extinction. Yet, however small this probability sometimes is, sooner or later species will die out because of external environmental changes. Arguments corroborating this have been given in genetical, ecological, and palaeontological contexts. The niche-width variation hypothesis relates genetic variation with a species' local distribution among habitats and with range size. This idea is also implicit in the ecological theory of risk spreading (Chapter 11), suggesting that the more variable a species and its environment, the more its extinction will be delayed. The operation of these processes, however, is difficult to observe, as correlations between a species' variability and that of its environment do not reveal whether the former is the cause or the effect of the latter, and restricted species can be rather variable genetically.

The maturation process in polyploid complexes in plants can show another relationship between the degree to which a species matches environmental variation. Stebbins (1971) observed that the older such complexes become, the more restricted their diploid members are, eventually dying out. On their side, the polyploid members, usually having a wider ecological amplitude and a larger geographical range, are better buffered against environmental fluctuations due to glaciation, mountain building and degradation, as well as to short-term climatic fluctuations. Polyploids would particularly occur in changeable and disturbed habitats, or as ruderals and weeds in cultivated areas and may extend their gene pool and range by taking up genes from diploids, thus becoming so-called compilospecies. Stebbins (1971) recognized five maturation stages, initial, young, mature, declining, and relictual, after which the complex eventually becomes extinct.

The latter arguments pertain to an individual's response to environmental variation, although they are usually clustered on various spatial scales into populations in the ecological sense, or into population genetic demes. Sometimes only sections of populations or demes die out, at other times they are exterminated completely, and in still other cases all populations in large parts of the range die out. The probability of regional exterminations depends on the number of localities occupied (Figure 52) (Hanski, 1982), on abundance, tolerance range, and range size. Mollusc species from the tidal zone, living up to one metre deep, have relatively higher local densities and larger geographical ranges, implying that the ecologically more variable species have the lowest extinction rate (Jackson, 1974). This accords with the lengths of geological periods from which fossil mollusc species are known, after categoriz-

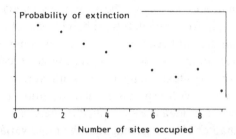

Figure 52. Extinction probabilities as a function of the number of locations occupied by insects found on mangrove islands (after Hanski, 1982).

ing them similarly to the recent ones. Moreover, species living in the zone of one metre or less have a greater dispersal efficiency than those from 1–200 metres. Although random fluctuations also occur at greater depths causing extinction, species in the higher, tidal zone are better adjusted to catastrophes such as extremely low tides, temperature or salinity fluctuation, or storms, possibly explaining their lower extinction rate.

Finally, palaeontological evidence indicates a negative relationship between range size and extinction rate or probability. Thus, a linear relationship exists between the logarithm of subtaxa number within a taxon and time, implying that species have an expected half-life of survival. Many authors objected to this statistical explanation of extinction, arguing that history shows distinct periods of mass extinction, alternating with periods of stasis or normal extinction (e.g. Jablonski, 1986; Raup and Jablonski, 1986; Stanley, 1986). In periods of normal extinction, biological properties apart from range size affect the probability of dying out, whereas in periods of mass extinction, due to catastrophic events, only range size matters, the larger ranges having the smallest extinction probability. From a biogeographical viewpoint both types are equally interesting, though for different reasons.

Optimum surfaces and individualistic spatial behaviour

Present-day communities are unique, ephemeral species assemblages, as each species behaves individualistically, each having particular source areas, and migration directions and rates particular for each species (e.g. Cushing, 1982; Cushing and Dickson, 1976; Davis, 1981*a*; Graham, 1986; Huntley and Birks, 1983; Walker and Flenley, 1979). Thus, species occur in one particular refugium depending either on local conditions or on chance processes and they migrate at rates depending on

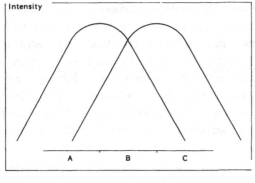

Figure 53. Hypothetical optimum responses of two species relative to the same environmental factor. With increasing intensities of this factor, there are parts where the species behave similarly, i.e. either increase (A) or decrease (C) in concert, or where they respond the opposite way (B).

specific dispersal capacities. But how can they migrate in different, sometimes opposite, directions under the same changing conditions?

Two-dimensional optimum surfaces result from a response to, for example, humidity and temperature, which are specific for each species in several statistics. These statistics are, for example, mean, standard deviation, and skewness, and jointly characterize the response to individual variables and their covariance. One explanation is that species differ in sensitivity to the same environmental factors. Figure 53 gives another explanation, showing lines that indicate the path followed by two hypothetical, ecologically different species, resulting from the same change in both humidity and temperature. These paths are relative to the species' response surfaces, their directions in these figures indicating numerical increase or decrease. These numerical responses differ, both for many parts of one optimum surface and for several different ones. Different physiological responses thus explain the individual, sometimes opposing reactions of species relative to the same environmental changes. In addition to these differences, owing to specific optimum surfaces, differences due to anatomy, morphology, life history, seasonality, or dispersal capacity are also superimposed. Moreover, species differ in source area, topography, and distances to be covered, which, taken together, make a species' spatial behaviour unique at any instant in time and explain why community composition changes continuously (cf. Gleason, 1939).

Conclusion: the range as a process

The concept of species ranges as optimum surfaces can be extended to that of a dynamic structure, expressing a continuous response to changing environmental conditions. Changes of and within ranges are not isolated incidents, but events exhibiting the very nature of species ranges. Climate as a principal component of a species' environment is a global, physical process to which species are permanently adapting. Species can adapt quantitatively to this variation in space by means of variation of local intensities, and qualitatively, by means of geographical differences in genetic composition. The speed of adaptation varies between species, and occasionally falls short, causing species extinction. Thus, similar to range shifts, extensions or contractions, extinction rates express the dynamics of both species and environment. Since both the species and environmental variation on all different scales are stochastic to a high degree, extinction should be described in stochastic terms as well. Species distributions are dynamic entities over long or short time scales and can be considered as outcomes of a process of perpetual adaptation to changeable conditions.

11

Population dynamic theories

Factors causing variation in broad-scale patterns and processes are not confined to those scales. Together with other factors, they determine the level of variation on other scales. Thus, questions about whether or not abiotic or biotic factors determine the size, location, shape, and dynamics of ranges, can miss the point, as both types operate. Fine-scale ecological studies can reveal how processes related to broad-scale levels of variation operate locally and how these, conversely, interact with those related to fine-scale variation. Their interaction can result in local or geographical differences in fine-scale processes within species ranges, thus explaining mechanisms of discordant variation.

Because broad-scale processes are principally the same as fine-scale ones and because they are intertwined, it makes no sense to distinguish them as geographical and ecological factors, respectively. Rather, such a distinction does harm to our understanding of spatial processes as such and at each level separately. Moreover, population dynamic methodology can be generalized to broad-scale processes; questions such as why population density and size do not increase indefinitely also hold for species ranges. Methodologically they are exactly the same.

First, two competing population dynamic theories will be described, population regulation by feed-back mechanisms and stabilization by risk spreading. Their common assumption of the existence of an economy or balance of nature will be analysed in a later section. The following two sections evaluate both theories, followed by a discussion of energy budgets at the individual level, supplementing demographic processes.

Population control versus risk spreading

On human time scales many populations are relatively stable; usually they do not expand to outbreak proportions, nor do they often die out. Part of this realization, however, may be due to a shortage of long time-series, or when they exist, to their pertaining to relatively common species. One possibility for explaining this apparent stability is that several species-specific feed-back mechanisms keep population densities mutually within bounds. The larger the deviation from some preferred density, the more intense the counteracting forces. For example, the higher the density, the greater the demand on the resources and the greater competition for this resource will be, thus counteracting further growth. The opposite may occur at low densities. A variant of this theory is that, due to interactive factors, food varies in quality rather than quantity. This mechanism is called regulation by means of population control.

Another mechanism for keeping population densities stable occurs when many unspecific and uncorrelated variables affect a population. The greater the number of variables, the smaller the variance of the fluctuations and the smaller the probabiliy of reaching outbreak proportions or of dying out and hence, the more stable the population. Normally, populations are subjected to many variables in their environment, but this number increases even more for large phenetic variation. Losses due to certain factors may thus be compensated for by gains resulting from others, at the same time and site or in others, or in the same stage of their life cycles or in others, for example. As this process would be the same as when big firms ensure their profits by investing in various economic sectors, the term spreading the risk, characterizing that process, was adopted in population dynamics (Den Boer, 1968; cf. also Andrewartha and Birch, 1984).

Thus, population control assumes ecologically functional demographic responses stabilizing population density and size, and risk spreading a non-biological, stochastic process.

The balance of nature
Both theories assume population processes to be spatially and temporally stationary. This assumption concerns the more basic theory of the balance of nature, adopted in the earliest stages of ecological history (Egerton, 1973). The theory of risk spreading gives an alternative mechanism of the stabilizing process, but maintains the assumption that populations are stable (Hengeveld, 1989*b*).

In the course of history, various biological processes have been used as criteria for proving or disproving the existence of a balance of nature. Thus, various, often opposing, theories are combinations of several of these criteria, resulting in a virtually unlimited number of theories

(Hengeveld, 1987*b*; Wiens, 1984). Characteristic for the balance of nature theory is that most criteria are demographical, the relevance of which is attacked by opponents, who usually do not give alternative mechanisms, nor attack the basic stability assumption. The theory of spreading the risk did the first, but it failed in the second.

Population dynamic models are usually untestable because most existing time-series are too short or lack sufficient detail. This untestability applies even more to risk spreading, since it is difficult, if not impossible to estimate whether a particular variance is due to a certain biological factor, such as fertility, to several unspecific ones, or to both. It is equally difficult to estimate from field observations that a small variance or fluctuation has effectively increased the expected time to a population's extinction or not. Stability is an inherent property of population dynamic models, expressed by the assumption of spatio-temporal stationarity concerning both the mean level and the variance of fluctuation. As this assumption is wrong, we need other models with other basic properties. Although the properties as such cannot be tested, the assumptions can and through them the applicability of certain models can be evaluated.

Because of the more or less regular geographical build up of species intensities over the range, variation between populations in space is not stationary, although implicitly it is assumed to be. This means that intensity levels are determined differently among different locations or regions (Andrewartha and Birch, 1954). We cannot think about those levels, or model them, in general terms, as if they were spatially stationary. In all cases with horse-shoe-like scatters resulting from PCA ordinations, we are dealing with a spatially non-stationary process; models assuming stationarity are applicable at single points on this curve only, that is for the particular conditions at single locations or regions, which are mutually incomparable at the level of variation of the factor concerned. The same holds for the variance of temporal fluctuation.

Despite the basic assumption of temporal stationarity in population dynamics, processes are also non-stationary in time. Davis (1986), for example, concluded from comparing temperature curves for several period lengths that at the scale of the last 1·5–2 million years of the Pleistocene, temperatures are stationary at the scale of 100000 years only, thereby reaching a different mean level from the present one. But Pleistocene temperatures are only the last phase of a cooling trend that had already started in the Cretaceous, and that will probably continue during the next millions of years. Temporal variation is only stationary for short periods, and is not comparable with that of other periods, resulting in the fact that present vegetations have no past analogues. Models based on stationarity assumptions obviously are of limited application, the subject of both ecology and biogeography being dynamic in essence.

Assumptions as to community stability are thus of limited applicability, also because virtually no species has the same spatial distribution and local dispersion, nor the same temporal pattern of fluctuation, resulting from dissimilar responses to the same factors, even when operating at the same scales. But population processes operate at specific spatio-temporal scales of variation. Consequently, prey–predator models are inapplicable to, for example, titmice and caterpillars responding with different rates to different factors operating at different scales. The theory of spreading the risk assuming non-specificity of interrelationships might have cured this, but as simulation results show (Reddingius and Den Boer, 1970), the mean fluctuation level varies independently of the variance; small variances may be accompanied by gradually increasing or decreasing densities, in fact illustrating processes occurring on different temporal scales. Moreover, many environmental factors are not independent, but covary or even interact in time and space. Similar to regulation theory, risk spreading does not explain different levels of variation over a species' range, nor in time.

Restriction to small spatial and temporal scales resembles taxonomic classification several decades ago, typified as a typological, non-dimensional approach (e.g. Mayr, 1957) in contradistinction to the multi-dimensional approach. In typological approaches some norm, essence, or type is assumed, the actual observations being considered as deviations from this norm. Simberloff (1980) analysed ecological concepts and processes from this viewpoint, concluding that this approach prevails in present-day ecological theorizing. Non-dimensional approaches are applied by local naturalists, ignoring spatial and temporal variation or dimensions, which equates to assuming them to be stationary. Multidimensional approaches include spatio-temporal variation and accounts for the properties of individuals or individual data points. According to the latter approach each point should be considered individually and explained in terms of instantaneous conditions and previous population levels (Hengeveld, 1988a, 1989b).

However, matters of scale are more complicated; processes operating at various levels of variation should not only be distinguished, but their effects should be integrated. For example, explaining optimum responses to temperature, Gause (1932) suggests that each temperature value

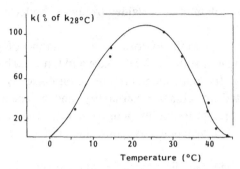

Figure 54. The logistic asymptotic levels K of the yeast *Saccharomyces cerevisiae* as a function of temperature (after Gause, 1932).

determines the intensity of ecological interaction by determining K, the asymptotic level of the logistic growth curve (Figure 54). Similarly, Beauchamp and Ullyott (1932) show that the outcome from competition between two planarian species depends on water temperature, *Planaria montenegrina* being more successful at temperatures below 13–14°C and *P. gonocephala* at higher temperatures. However, other factors may also be involved; at temperatures of about 9°C, *P. alpina* is more successful in fast-streaming water, but *Polycelis cornuta* dominates in slow streams. Recently, Jerling (1985) showed for *Plantago maritima* that population dynamic parameters vary along a 60 metre gradient from a waterfront to a meadow. Near the waterfront population turnover is high because of high fertility and seeding rates but, due to flooding, low life expectancy. Away from this front, population turnover decreases because of increased life expectancy, despite poorer fertility, seeding and establishment. The genetic fitness of this species appears to be differentiated along this gradient as well (Jerling, 1988). Similar variation in time and space may occur over a species range, accounting for variation in turnover rates due to number of generations per year and over the year, number of predators, parasites, and diseases, or climatic fluctuation.

All this constitutes the multidimensionality of ecological species behaviour and adaptation and shows the limited applicability of population dynamic models based on stationarity assumptions. It is not surprising, therefore that, probably because of this, their predictive value is low under both natural and agricultural conditions even when applied to fine-scale processes. Biogeographical data are crucial by showing the broad-scale spatial non-stationarity of these processes, similar to Quaternary research showing long-term temporal non-stationarity.

Extinction probabilities explained by energy budgets and spatial dynamics

It should be possible to formulate an outline of population dynamics explaining non-stationarity in space and time and based on the species' properties. Its main elements are physiologically based risk distributions related to climatic variability and a species' spatial dynamics. The first relate to energy budgets as principal determinants, and the second to ecological patch dynamics and epidemiological spatial diffusion processes.

Physiologically, one can distinguish between several energy flows, those for feeding, storage, growth, reproduction, activity, and the organism's basal metabolism. Given these flows as parameters, one can, in principle, model energy use for various species with different sizes or population dynamic properties (Kooijman, 1989; cf. also Blem, 1973; Kendeigh, 1976; Currie and Paquin, 1987; Turner *et al.*, 1988; Wright, 1983). Spatio-temporal frequency distributions of required temperatures thus greatly determine the probabilities of local extinction or survival (Neilson and Wullstein, 1983) and, hence, a species' geographical parameters (e.g. Koskimies and Lahti, 1964). Indirectly, energy can be limiting because of limited foraging time (e.g. Alkon and Saltz, 1988). Apart from responding physiologically, species can respond to environmental variation by shifting their intensity distribution in space.

Parry and Carter (1985) (see Chapter 12) estimated the frequency of occurrence of the minimum heat sum a species requires from a 323-year time-series, obtaining the risk a species, oats, runs of not completing its life cycle, by not producing viable seed within a year. Thus, they obtained explicit estimates of the risk distribution of species for several altitudinal levels, which can similarly be applied to different parts of species ranges. Wigley (1985) pointed out that risk increases non-linearly towards distributional margins and is steeper for more marginal parts and for two years in succession than for single years by a shift of frequency distributions (Figure 55). Thus, yield variability of several crops is lowest under optimum conditions, increasing towards physiologically and geographically marginal conditions, indicating increased risk (Figure 55) (e.g. Klages, 1942). Since extinction risks concern temperature sums over particular periods, even small differences in mean temperature have great effects, depending on period length and steepness of temperature increase at the start and end of the growing season. Though correlating with minimum and maximum temperature, they give better insight into

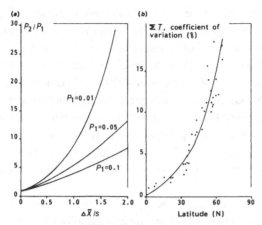

Figure 55 (*a*) Expected change in the probability of occurrence of events with initial probabilities P_1 of 0·1, 0·05, and 0·01 for shifts ($\triangle$) in location of the average $\bar{X}$ of a standard normal distribution of size $\triangle\bar{X}/s$ along a gradient (after Wigley, 1985). (*b*) Observed variation of heat supply for rice in Japan towards its northern range limit (after Uchijima, 1976).

an important, although not exclusive, mechanism of range structure and limitation.

Temporal risk distributions can explain a species' range structure and delimitation, whereas their shifts necessitate species to migrate either within their range, or to invade other regions. On broader time scales, evolutionary adaptations can be superimposed (Hill, Read and Busby, 1988). Here, I discuss aspects of both internal range dynamics and invasions outside the range.

Black (1966) found for measles that below a certain threshold value of the number of susceptibles, the epidemic fades out; the larger the host density, the smaller the distance of susceptibles, the longer the duration of the epidemic, until the occasionally occurring epidemic is continuously present as an endemic. Cliff *et al.* (1981) extended this, distinguishing three types of epidemics, those with (1) regular, continuous waves, (2) regular, but discontinuous waves, and (3) irregular, discontinuous waves for populations of >650000, >10000, and <10000 individuals, respectively. Type 2 represents fading out, direct overflows from epidemics elsewhere, and type 3 occasional infections of (partly) susceptible populations. The temporal disease pattern of smaller communities thus depends on the number of contacts, being itself a function of its size, distance to endemic centre, and transfer efficiency. This transfer efficiency can vary in time, depending on various factors. For example, from the turn of the century onwards, Reykjavik experienced a type 2 epidemic, the remainder of the population in the course of time developing from a type 3 to a type 2 epidemic, due to increase in population, traffic intensity, and several other social developments. Also, generally warmer conditions during the 40 years after 1920 resulted in less drift ice, which may have affected epidemic incidence through blocking harbours. This process, complicated by expansion of local population centres will have wider application than in epidemiology only; similar situations will occur from the range centre to the margins and under changing ecological conditions as a spatial component of the dynamics within and between populations.

When conditions change, species can respond by either contracting or expanding their range, depending on epidemiological parameters. Thus, range contraction depends on the individual's expected longevity determining Carter and Prince's (1981) removal rate, which is high for annuals and low for perennials, particularly trees. Great longevity either postpones extinction or reduces the chance of extinction through allowing to bridge unfavourable periods. After its attack by *Endothia parasitica*, the American chestnut still survives, though hardly ever reproducing generatively, if at all. Another consequence of environmental change can be that the species' dispersal capacity falls short because its spatial transfer probabilities do not match the probability distribution of the pattern of its habitats. This is similar to the temporal mismatch when favourable periods are so infrequent that both the organism and, for example, its seed bank becomes exhausted. As the locations of these distributions vary continuously, extinction can be regarded as a process rather than as an event, depending on demographic and other species-specific parameters.

Invasion processes, depending on the same parameters and on migrational ones, are even more complicated. Essentially, these parameters are the same as those relevant in the internal dynamics of ranges and populations; they are more apparent because of the species' arrival in a new region. The recent invasion of the Collared Dove, *Streptopelia decaocto*, into Europe is a case in point (Hengeveld, 1988b). This invasion can be described in terms of diffusion, plus a component of population growth (e.g. Skellam, 1951) (compare Kareiva and Odell, 1987, for local interactive processes). Its diffusion is stratified, resulting from short-distance neighbourhood diffusion plus that of long jumps of up to several hundreds of kilometres. From the new bridgeheads away from the invasion front, the species usually spread short distances in various directions. The spruce budworm offers a variant on the invasion process, where immigrants may lift an endemic level of local variation to another, higher level, resulting in a process that can lead to outbreak proportions according to a process known from catastrophe theory (Clark *et al.*, 1978). If this applies, we need spatially interdependent population models for understanding local outbursts rather than the spatially confined ones used at present. This would raise population dynamics from the dynamics of single populations to that of the species as a whole.

Conclusions

Population dynamic processes are neither quantitatively stationary in time and space, nor qualitatively concerning the parameters involved. Therefore, present-day population dynamic models ultimately based on the balance of nature theory assuming qualitative and quantitative stationarity do not apply when put into broader-scale contexts. Also, interpretations based on biotic or abiotic variables are not alternative to or excluding each other, but are integrated through interactive processes.

Spatial intensity of trends in ecological variables in dynamic environments can be described in terms of frequency distributions. As these distributions are themselves constituted by temporal distributions on a finer scale and constitute other ones on broader scales, spatio-temporal processes should be described in stochastic terms. Similarly, the process of dispersing from one location to another, either resulting in stable spatial dispersions or in migration, is essentially stochastic as well, only under certain conditions reaching deterministic predictability. Basic to these processes, however, are the energy requirements of individual organisms described by direct input from their environment or indirect input through feeding relative to various forms of energy loss. The whole process can be formulated in terms of risk theory, where individual data points are considered separately rather than as deviations with a particular variance from their mean.

Thus, ecological processes illuminate the dynamics of species ranges and, conversely, long-term geographic processes illuminate the nature and variability of ecological ones.

Summary of Part III

The material discussed in Parts I, II and III has been arranged from very broad-scale to fine-scale patterns of variation. Moreover, Part I discussed phenomena in which neither the identity of taxa, nor their specific, biological properties are of prime concern. In Part II only these biological properties were discussed irrespective of taxon identity, and Part III concerned processes in identified taxa with their specific properties. The chapters in Part III were arranged similarly, Chapter 9 discussing species ranges as static entities, neglecting temporal parameters, whereas Chapter 10 included temporal variation. Finally, Chapter 11 considered the fine-scale spatio-temporal dynamics of local populations put within the framework of the range as a whole. Throughout, all phenomena were described statistically and, accordingly, explained in stochastic terms.

The arrangement from broad-scale to fine-scale phenomena can easily lead to introducing reductionism into biogeography. As, in practice, reductionism involves causal explanation with phenomena occurring on finer-scale levels without referring back to the phenomenon at the original level of variation, this pitfall can be avoided. Part IV therefore attempts to close the methodological circle by explaining broad-scale variation as discussed in Part I by phenomena described in Parts II and III.

One can also easily be trapped by the statistical formulation of the phenomena concerned. Then, one investigates phenomena of particular taxa such as species without considering their identity and biological properties. This is usually done in both vicariance biogeography and in MacArthur and Wilson's (1967) equilibrium theory. It can also be recognized in present-day discussions of topographical, geological and Quaternary mass extinctions. In all these cases, explanations are formulated in non-biological terms, that is in geological or in extraterrestrial terms, or in those involving catastrophes. But adding more biological information can make such factors unnecessary. Guthrie's (1984) analysis as opposed to Martin's (1967) original overkill hypothesis for Quaternary extinctions is an example of this. As the broad-scale, biologically unspecific biogeographical patterns of concordance of Part I, though statistically defined, are also biologically caused, this is another argument for describing them in terms of the biological processes treated in Parts II and III.

Important aspects of Part III are the spatial (Chapter 9) and temporal (Chapter 10) non-stationarity of species intensity and the differing processes and causes (Chapter 11) that result in this non-stationarity. Species behave dynamically and individualistically in space, communities or biogeographical concordance patterns being temporal aggregations without antecedents and varying in composition from one locality to another. As climate is also dynamic and varies spatially, its various elements to smaller or larger degrees being independent, present-day distribution patterns will often tell little, if anything, about patterns in the past. Also, hypotheses as to the nature of ecological or biogeographical spatial units as co-evolved interactive processes should be viewed with caution. Present-day patterns of interaction are by no means always the key to the past. On the contrary, interactions may be of secondary importance relative to climatic factors in explaining distribution patterns of species, the latter factors often being conditional to the first. And this is the point where the two aspects of spatial analysis discussed in this part of the book, biogeography and ecology, merge.

IV

Species ranges and patterns of concordance

In the previous parts, many biogeographical phenomena were arranged from coarse, broad-scale patterns to specific, fine-scale processes defining risk distributions of individual species. The three principal factors determining survival probability of a species were (1) its energy budget, (2) its spatio-temporal dynamics as a response to environmental variation, and (3) the difference between (1) and (2), representing the time-lag in spatio-temporal population dynamic responses relative to this environmental variation and known as 'the historical factor'. In this part I will integrate information obtained from analysis according to these three factors and at all scales.

This integration can be achieved from two viewpoints, namely discontinuous variation in space and time, and stochastic variation on various scales. The first should give answers to how individualistically behaving species can form recognizable units on various scales, despite their often contrasting responses to regional environmental conditions. It also considers representativity of present-day assemblages of taxa, and, hence, the nature and duration of species interactions inferred from those presently observed. As responses to abiotic and biotic conditions are mainly thought of as deterministic processes often disturbed or directed by random environmental processes, I will also consider how far patterns of variation exist in apparently chaotically varying factors. If taxa reflect broad-scale environmental patterns, they may still be subjected to random variation on finer scales and vice versa. The first of these two possibilities connects broad-scale patterns with local risk distributions of individual species and the second, fine-scale assemblages with broad-scale spatial and temporal discordance of species co-occurrences. Even if the ideas presented here seem attractive, they must be tested and extended further, the implications of this form the basis for Chapter 13.

12

Discontinuous variation in space and time

The abundance distribution within species ranges is unevenly spaced and varies over several scales. Spatio-temporal variation is not continuous, but can be spatially patchy and explosive in time. Because weather develops stochastically, local conditions are unpredictable and species behaviour is erratic.

For example, the range of the Desert Locust is well-defined, as are regional abundance centres and its seasonal spatial variation within this range. Moreover, particularly Rainey (1963) and Waloff (1966) showed implicitly that this spatio-temporal variation is stochastic, both in the short-term and over many years. This stochasticity results from shifts in both the prevailing wind pattern north of the Intertropical Discontinuity and in the Discontinuity itself. Yet, this prominent wind system varying on several spatio-temporal scales is not the only operative factor. Rainey, Betts and Lumley (1979) described the decline of outbreak frequency in the 1960s, which coincides with a global climatic change, starting in the 1950s (Lamb, 1979). Moreover, from 1908 to 1966 four major outbreaks occurred, whereas in the same period two related species, the Migratory Locust (*Locusta migratoria migratorioides*) and the Red Locust (*Nomadacris septemfasciata*), each showed only a single outbreak period (Waloff, 1966), a difference attributed to the greater mobility of low-density populations of the Desert Locust relative to the other species. Thus, its spatio-temporal variation results from several causes operating on different scales. The response to environmental variation is particular to the species, resulting from its dispersal efficiency and several population parameters, such as parasitism and cannibalism.

It seems that, unless spatio-temporal variation is discontinuous, species vary randomly in space. It can be hypothesized that we have to explain the occurrence of sharp biotic delimitations, ecotones or tension zones, when these are found at a particular scale. But when this variation is regular, one may wonder if there is one single relevant interval, or if geographical phenomena vary on several, superimposed discontinuous scales. Also, variation can be regular and biologically interpretable at one scale, and random at another.

Biogeographic provinces and their dynamics

From knowledge of ecological preferences of oceanic algae, Van den Hoek (1982*a*, *b*) and Yarish, Breeman and Van den Hoek (1986) ascribed the similarity in species distribution to similarity in temperature preference and thus to species' physiology. But are processes in terrestrial species similar to marine ones and do similar discontinuities to those caused by ocean currents occur? Flow of air masses indicates that on land geographical patterning may also be discontinuous, being caused by atmospheric currents similar to those in the oceans.

For example, storm tracks over North America as indicators of airmass movements explain the location and timing of outbreaks of two insect species in eastern Canada (Chapter 10). Here, three regions of air mass origin can be distinguished: polar regions, continental regions, and maritime regions. Air masses from these regions keep their physical properties and carry these across vast continental stretches. Thus, Wellington (1952) characterized areas of high outbreak incidence by a minimum frequency of air masses arriving from one of these areas during certain sets of years. More recently, Neilson and Wullstein (1983) and Neilson (1986) explained the present range location of two oak species, *Quercus gambelii* and *Q. turbinella*, in southwestern North America in a similar way. They transplanted seedlings of *Q. gambelii* along both several microhabitat gradients and gradients of air masses to determine seedling mortality caused by spatially discontinuous climatic variation. They also did experiments in growth chambers on the drought physiology of *Q. turbinella* seedlings. From this information they showed that the northern ecotones are primarily caused by spring freeze stress due to polar air, and by summer moisture stress due to drought from the Arizona monsoon. Spatial shifts in the seasonality of these two air mass gradients over time explain the biogeographical history of these species during the present interglacial. Similarly, Borchert (1950, 1971) showed that the North American grassland biome coincides with the outflow of Pacific air across the Rocky Mountains (Figure 56). As it loses its water vapour at higher altitudes in the mountains, this air is relatively dry. However, its relative frequency of occurrence diminishes gradually eastwards, resulting in increasingly more humid regions in eastern North America. Thus,

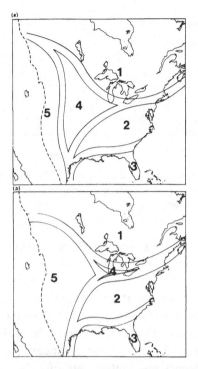

Figure 56. Locational variability of North American climatic regions. Regions are characterized by snowy winters (1), rainy winters (2), rainy tropical summers (3), relatively dry winters and occasional dry summers (4), and relatively dry winters and summers (5). Figure 56 *a*: reconstructed regionalization for the late post-glacial; Figure 56 *b* recent (after Borchert, 1950).

in the west, adjacent to the mountains, short grasses occur, which are gradually replaced eastward by mid-grasses, tall-grasses, and forests respectively.

Elaborating upon this, Bryson (1966) defined many North American biomes in terms of air mass currents across the continent. To do this, he constructed frequency distributions of the intensities of several climatic elements for many North American weather stations. When these distributions were multimodal and showed a geographical pattern, he divided them up into air masses of different geographical origin (Figure 57). He then calculated the relative frequencies per locality for each unimodal distribution recognized within the multimodal ones. Because he used information on several weather elements, such as precipitation

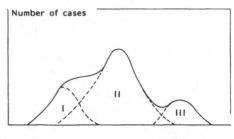

Figure 57. Schematic three-modal frequency distribution of local temperatures reflecting air currents coming in from three areas of origin (after Bryson, 1966).

and wind patterns, as well as that on global air currents, Bryson could classify the North American climate statistically in genetic and dynamic terms. This classification was correlated with existing vegetation classification of biomes, allowing their description in terms of dominant weather types or their relative frequencies (Bryson and Hare, 1974). Thus, the North American tundra zone coincides with the region where arctic air predominates in 10–12 months annually.

But annual frequency distributions are often too coarse for species with a pronounced seasonal occurrence. The same technique was therefore applied to monthly data, defining regional climate in terms of seasonal trends in relative air mass frequency of different origin. Now the southern tundra border coincides with the southern fringe of arctic air in summer and the southern border of boreal forests with arctic air in winter (Figure 58). Comparable to the oak species, the Corn Belt lies between the southern fringe of polar air in spring and the eastern fringe of Pacific air in winter. Similar correlations exist for other biomes, such as the southeastern deciduous forests or the Central American prairies. Thus, discontinuous spatial variation in floristic composition can be correlated with discontinuous climatic variation due to seasonal air mass dynamics. In their turn, movements of air masses result from physical processes in the global atmosphere, caused by differential solar heating of the earth, its daily rotation, and the alternate heating of the Northern and Southern Hemispheres during the year.

Krebs and Barry (1970) tested this explanation by predicting the same correlation for the arctic and alpine tundra, the forest tundra ecotone, and the taiga in Siberia. As in Bryson (1966), they chose the location of

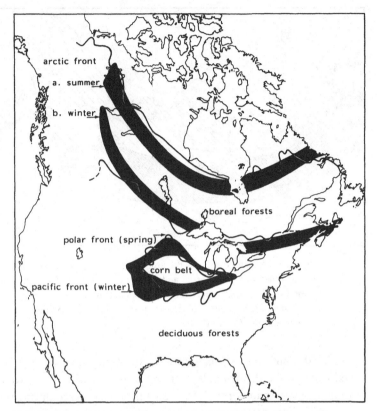

Figure 58. Coincidence of average positions of fronts of air from various origins for different parts of the year with several North American biomes and regions of crop production (after Bryson, 1966).

the arctic front in July estimated for the period 1952–6. For each year separately, they calculated the daily median location of this front and expressed the year-to-year variation of this median in terms of quartiles and deciles (Figure 59). In accordance with Bryson's (1966) hypothesis for North America, the Siberian tundra–taiga boundary correlates closely on this scale with the median location of the arctic summer front.

The correlation between vegetational composition and climate can also be substantiated by vegetational response to changes in climatic patterns. For example, we can look at possible effects on range delimitation and vegetation composition of the North American prairie due to the climatically deviating period of the 1930s and 1940s. At that time Pacific air penetrated much farther to the east than usual, expressed by, for

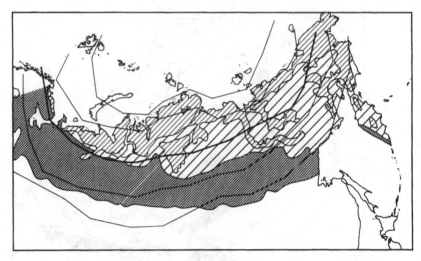

Figure 59. Median position of the arctic front across Siberia (heavy drawn line) together with the quartile and decile occurrences of this front (thin and thinnest drawn lines, respectively). Biomes are shown by hatching, intermediate: tundra and alpine; thin: forest-tundra; heavy: boreal forest (after Krebs and Barry, 1970).

example, the amount of precipitation as a percentage of that in other years, causing the dry triangle east of the Rocky Mountains to stretch farther eastward than normal. The vegetation responded during the next years, the short-grasses eastwards replacing the mid-grasses, these replacing the tall-grasses, and these in their turn the forests at the eastern fringe of the prairie zone. The same happened during the late post-glacial as compared to the present (Borchert, 1950; see Mitchell, 1979, for a 22-year drought cyclicity). The opposite occurred during the cooler and more humid 1950s and 1960s (cf. Coupland, 1974). Similar shifts occurred in the southern Canadian tundra, which was replaced during milder periods than present about 3500 and 900 years ago (Bryson, Irving and Larson, 1965).

Apart from broad, vegetational shifts, the species responded individu-alistically to climatic changes, rather than as community members. Their responses are determined by specific physiological, anatomical, and morphological properties, resulting in dramatic shifts in species composi-tion. For example, the grass *Bouteloua gracilis* withstood periods of great drought better than *Buchloe dactyloides*, because it has rhizomes instead of stolons, as in the latter species. But because of this morphological trait, after favourable periods *Buchloe* can build up its populations quicker than *Bouteloua* (Coupland, 1958, 1959, 1974). In general, deep rooting species were, at first, at an advantage relative to those with shallow roots, but as drought continued, this short-term favourable property turned into a disadvantage when survival depended on occasional, short showers. Also, species occurring early in the year were less affected by the general drought than species occurring in summer.

The implication of these individualistic responses to ever changing ecological conditions is obvious; on a short historical time scale and that of the period since the last glaciation, species are independent and do not form stable, structured communities. Also, vegetational composition was different during various glacials and interglacials (cf. Birks, 1986). Moreover, interglacials such as the present one are climatically deviant periods of short duration, present-day vegetational composition having no analogue in former historical or geological periods.

This ever-lasting continental shuffling and reshuffling of species, forming spatially and temporally unique combinations, has two consequences for biogeography. First, it is difficult, if possible, to reconstruct vegetational composition for times longer than 8000–9000 years ago and, conversely, to reconstruct climate from differences in those compositions. Climatic reconstruction therefore depends on knowledge of the combined autecologies of species, rather than on that of community ecology. Second, it is difficult, if possible, to reconstruct processes in the past from present-day geographical patterns of species distributions; too much has happened to biotas over the millions of years that usually separate the past from the present.

Thus, not only can biotas be classified, but one can also explain spatial and temporal discontinuities, even when the explanation concerns fluid atmospheric systems. Yet, on a finer spatial scale or a broader temporal one, the dynamics of the categories is too great to consider them biological entities. Species coexist under certain conditions, but retain their individuality. Biogeographic units are temporary and geographically fluid spatio-temporal patterns of overlap because of a certain degree of concordance of species responses to local and temporary overlapping environmental conditions.

Glacials and interglacials

Similar to spatial variation, temporal variation of environmental conditions can be continuous or discontinuous. For example, at present, at least 16 glacial periods during the Pleistocene can be shown to have existed. This indicates that climate fluctuated greatly during the last two million years, whereby the Quaternary as a whole is a cold period following the warmer Tertiary. This poses the problem of how cold periods originate, how common and long-lasting are interglacials, how variable and fragile are atmospheric conditions, and what side-effects have changes in temperature, particularly on atmospheric aridity and direction and intensity of wind patterns, sea currents, and sea level. All this is significant to biotic response rates and the similarity of present-day biotic compositions to those in previous times.

From the synoptic climatic processes described in Chapter 10 it follows that changes in the global energy budget do not affect all local temperatures equally, but differentially; spatial patterns of climatic variation alter both in extent and location. Moreover, small temperature changes have large effects on the patterns of global heat exchange. Also, the amplitude of temperature fluctuations increases with latitude, which on top of stepwise changes in temperature exchange enhances the transition rate between glacials and interglacials. Another factor with similar effect is that, due to a change in climate, sea currents take a different course, thus reducing poleward heat transport. Even during a minor climatic variation of the Little Ice Age between 1550 and 1700, the Gulfstream headed east–southeast instead of northeastward as it does today. During the glacials it did not exist at all, leaving the Atlantic north of 40°N covered with pack ice (Figure 60) (e.g. Ruddiman and McIntyre, 1981).

But ice ages involve more than temporally extended polar ice caps. Rather, they involve dramatic global and regional rearrangements of many climatic elements, affecting all environments differently. In fact, the ice covered only 9% of the earth at its maximal extension against 3% at present. Clearly, other parameters should be added if we are to understand changes in living conditions on regional and geographical scales.

Temperature changes are paralleled by global aridity changes; the colder, the more arid the climate becomes, present-day warm interglacial conditions are also among the wettest. Because of this, sea level lowers, being enhanced by isostatic crustal movements in many parts of the world. In turn, this causes, for example, great parts of the Sunda Shelf and the Sahul Shelf to run dry. This, again, blocks ocean currents through the Torres Strait, the South China Sea, and the sea south of Mindanao, removing their buffering effect and thus making the region more continental (cf. Nix and Kalma, 1972). Seasonal climates will have been more

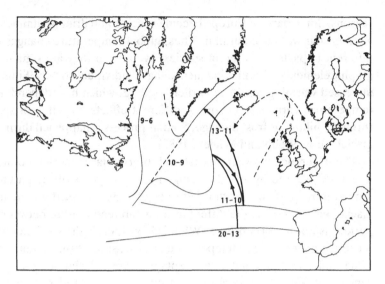

Figure 60. Position of the front of pack ice in the North Atlantic for various periods after the last glacial. The periods are indicated in thousands of years (after Ruddiman and McIntyre, 1981).

extensive during the glacials and the perhumid climate more restricted (Whitmore, 1981). Biogeographically, this joined the seasonal zones north and south of the equator, and dry areas extended. Africa was drier than at present, areas once connected being separated more recently. Similarly, Amazonian forests were reduced to several forest islands, serving as refuges where populations would have followed their own, independent evolutionary paths (e.g. Haffer, 1969). But we should be cautious with the latter interpretation, as calculations on bird distributions do not indicate that observed distribution patterns deviate from random expectation (Beven, Connor and Beven, 1984; Endler, 1982).

The glacials also differed in strength, direction, and timing of prevailing winds, altering seasonality in humidity and precipitation. Particularly the geography of seasonal precipitation altered. For example, certain regions of North America are characterized by winter precipitation and other regions nearby by summer rain. When seasonal air-flow patterns shift, large areas may experience both winter and summer rain, affecting species assemblages. For Western Europe, Lough, Wigley and Palutikof (1983) showed that small, global temperature increases result in spatially differential increases or decreases in temperature and precipitation. Moreover, these increases or decreases are dissimilar for the four seasons. Some parts of Europe become warmer and drier in summer, but warmer and wetter in autumn. Thus, global temperature changes have different impacts on different species, even when they are similar geographic elements, thus making them behave individualistically. Similarly, increased warming results in earlier budding, which has, depending on temperature variability in spring, different effects on different species through chilling or frost damage. Similar processes happened during the glacials (e.g. Webb and Wigley, 1985).

The difference between glacial and interglacial mean temperatures, about 5°C, caused by a slight change in the steepness of the latitudinal temperature gradient, is often felt to be too small to start and maintain glaciations. However, the difference in mean temperature between two periods is not important here. What is important is the total amount of incoming energy. Mean temperature is a measure of this amount only. Other important parameters for explaining both glaciations and species ranges are annual temperature and humidity amplitudes. Parameters such as these determine if a species receives just sufficient energy for its basal metabolism, temperature regulation, and growth or if it can also spend energy in reproduction. For migrants there should also be energy available for keeping up energy reserves during the journey (Blem, 1980). Similarly, plants require energy for their basal metabolism and reproduction, in their case energy losses from evaporation sometimes becoming critical, thus limiting the species' range. When its annual energy budget temporally falls short, a species dies out, either locally or completely. Depending on its seasonality and requirements of various stages in its life cycle, other fluctuation parameters come into play.

Thus, just as in space, at certain scales environmental factors vary discontinuously in time. And so do species, not gradually adapting to each other under constant or gradually changing conditions, but at one time quickly interspersing with other species and at other times separating out again and forming new combinations (Walker and Flenley, 1979).

The analysis of scales of variation
One of the techniques for separating effects of different sources of spatio-temporal variation is spectral analysis. It estimates the autocorrelation of a series of observations for lags of size n; when $n=0$, $r=1$, for $n>0$, r varies between -1 and $+1$. Mitchell (1976) applied this partly intuitively and partly quantitatively and distinguished several hierarchically arranged scales of variation, fine-scale variations being superimposed on broad-scale ones. Thus, day-to-day fluctuations due to

local thermal convective buoyancy are superimposed on longer-scale diurnal variations, these on synoptic eddies, caused by atmospheric inertia, etc. As the time scale of the processes concerned is logarithmic, their effects are geometrically related to their duration. Apart from this differential weighting according to their duration, processes can also be weighted according to their relative amount of variance, representing the intensity variation of climate. Furthermore, differences in peak height and their irregularity over time mean that fine-scale intensity variation is either smaller or larger than that on a broad scale; when it is smaller, the duration of the process is more important than its intensity. The spread around the modes expresses either variation in cycle length or observational uncertainty, the diurnal and annual cycle being the sharpest. Finally, Mitchell (1976) suggests that most climatic variation is stochastic, expressed by the area underneath his curve. The peaks only account for excess variation identified phenomenologically, implying that part of these identified sources of variation is also stochastic, only a small part of the total variation being deterministic. Thus, on a fine scale daily and annual variation are deterministic because of the earth's rotation around its axis and the sun, respectively. Similarly, the fortnightly variation, caused by the rotation of the moon around the earth is deterministic. But the synoptic atmospheric eddies at a slightly broader temporal scale are stochastic.

Yet another technique for distinguishing effects of several, superimposed sources of variation is harmonic analysis. For example, Horn and Bryson (1960) simulated seasonality in precipitation over the United States by defining several, superimposed sinusoids for the first, second, etc., order harmonics for various periods within the year. Thus, the first harmonic represents one annual precipitation period, the second harmonic two, etc. The phase angle of the sinusoid relative to a chosen amplitude its duration, which can be correlated with biological variables. The various local angles and amplitudes of the harmonics of different orders can be mapped individually, jointly, or as the ratio of local variances, etc. Several superimposed harmonics integrate the annual pattern of precipitation and their geographical distribution. Their analysis showed that the area west of the Rocky Mountains is characterized by winter rain, northwestern Washington receiving most in early January and southern California in early February. Occasionally this seasonally changes over short distances, as a sharp dividing line between two regimes lying, for example, in southern Utah and southern Nevada near Arizona.

Thus, not only the seasonal frequency distribution of occurrences of certain air masses explains the local presence or absence of species, but so does the seasonality of spatial interactions of these masses. Second, climatic conditions change considerably over short distances in, for example, seasonality of precipitation. Slight changes in the general atmospheric circulation alter local air flows and hence their pattern of interaction. This, in turn, alters a region once favourable to certain species into one hostile to them, or the reverse. Third, species pass the various phases of their life cycles at certain rates and have to adapt to local seasonalities in adjacent regions. If not, they have to migrate seasonally between areas with different temperature and precipitation regimes.

The integration of scales of variation

However clearcut results of our analyses may be, individual organisms are subjected to the total environmental variation, which, as seen through the species' eyes, is favourable or not, and always dynamic to some extent. Therefore, it is important to measure multiscale variation in its entirety apart from separately according to the level of fluctuation, the duration of favourable or unfavourable conditions, etc. The fractal dimension D of an area or a certain time period measures variability on several scales jointly (e.g. Burrough, 1981, 1983a, b, 1986) and can thus be used when no clear spatial or temporal scale of variation is apparent. Thus, unlike most statistical techniques, such as analysis of variance, multivariate analysis, harmonic analysis, and spectral analysis and their spatial counterparts that aim at discriminating between sources of variation, the fractal dimension D expresses the total spatio-temporal variation in one single figure. Graphically, D expresses the roughness of the curve expressing, for example, temporal fluctuations or spatial variation. Its value is high for great variability and low when the environment is temporally stable or spatially monotonous. For a uni-dimensional fractal function, such as temperature fluctuation over time, D varies between 1 for a straight line, depicting constant temperature, and 2 when the fluctuation curve fills up the two-dimensional graph paper completely. For two-dimensional surfaces, D varies between 2 and 3, and so on. Cubic fractal functions may represent, for example, the spatial variability of species intensities or their environments through time.

This measure makes two important assumptions: (1) the polynomial concerned is not differentiable, that is it cannot be split up into infinitely small straight lines, and (2) the scale of variation cannot be recognized from the observations, the total variation at any scale of resolution being identical. Polynomials of this type originate from sampling of probability distributions, each characterizing variation due to stochastic processes on n scales. Contrary to Brownian motion, in fractal variation the individual scales of variation are correlated. Areas may differ in the value of D or along transects, since, as a weighted sum of effects of several stochastic processes, the component processes usually differ. Thus, differences in D among areas, periods, or species can be tested according to their different level of compound variation. Essentially, fractal variation is the null model for unweighted, multiscale random variation (cf. Burrough, 1986, for further details and applications).

Global unity of climatic variation

Environmental variation is also spatially integrated, as it results from one single process, the differential heating of a rotating earth. On a global scale, several broad, latitudinal zones can be distinguished, each characterized by prevailing air streams. Near the equator there is a belt of easterly trade winds and near the poles a belt of westerlies, steered by meandering, high altitudinal jet streams. The subtropical zone lies between these wind systems, on the Northern Hemisphere characterized by clockwise whirling anticyclones and north of the circumpolar vortices the anticlockwise whirling cyclonal winds. The poleward heat transfer forms the force behind these wind systems. Depending on the solar heat input, the two opposite flows balance at a certain latitude, which is higher the greater the amount of incoming heat. If the heat input into the atmosphere is stable, the pattern of air flows is also more or less stable and coherent. However, this input varies both within and between years; within the year it varies because in winter the meteorological equator is south of the topographical equator and in summer it lies north of it. Although it shifts continuously, seasonal climatic variation is discontinuous according to the integer number of loops in the circumpolar vortex. As global climate does not adjust instantaneously to a new number, the former condition can be restored temporarily, resulting in a period of instability. For several years together, a number of natural seasons instead of the four on our calendar can thus be defined statistically, which vary regionally (e.g. Bradka, 1966). Thus, biological seasonality should match the onset and duration of these natural seasons. High-altitude air currents, and strata, add a longitudinal component to geographical shifts in the seasonal patterns of air masses and their flows. The onset of a particular season means different things to adjacent regions on the same latitude when in summer, for example, the loop number of the jet stream decreases and their meanders take other courses.

Blocking of summer temperature and precipitation patterning over Europe forms another aspect of within-year global climatic variation. This happens when, because of its size and low flow rate, a loop in the jet stream remains fixed for some time, rather than moving eastward as usual. When a blocked air flow affects Europe, it also affects the distribution of air masses over the Mediterranean and North Africa, thus causing moister conditions in the former and dryness in the Sahel. Apart from this, Atlantic surface temperatures also influence North African climate (Lough, 1980). Correlation of biogeographical phenomena with climatic processes on a European scale is therefore difficult; static populations adjacent to each other can be affected differently. Conversely, dynamic populations, such as those of migrant birds, are not only affected by African conditions, but also in their northern breeding grounds, which can be dependent.

Between-year variation shows similar global patterns. For example, a change in world climate during the 1960s was not uniform but regionally differentiated, both equatorial and subtropical anticyclonal zones having excessive rainfall and the intermediate zone being drier than normal. Contrary to these zonal changes, the mid-latitudes experienced alternating meridional regions of rainfall excess and deficiency, possibly due to a longitudinal shift in the main course of the meanders of the jet stream (Lamb, 1966). Thus, the general temperature decrease in the 1960s was accompanied by an equatorial shift of subtropical anticyclones and the subpolar depression zone and by a more meridional and shifting pattern of air flows in temperate regions. This can affect the living conditions of many species. Conversely, warmer summer conditions during the Hypsithermal at 6000 B.P. were azonal as well and heterogeneous from North to South Europe (e.g. Webb and Wigley, 1985), similar to temperature conditions in the present century (Lough *et al.*, 1983). Precipitation and the distribution of cloud cover shows a similar geographically differentiated patchiness which, as temperature, varies within the year (Henderson-Sellers, 1986).

Among other things, meridionality was also high during the glacials, indicating low jet stream velocities and blocking. Also, the centre of the circumpolar vortex occurred near the region of Baffin Island, and west and northwest Greenland. Because of meridionality, air with a high moisture content travelled over eastern North America and Western

Europe, thus bringing snow to Canada and Scandinavia and building up the Laurentian and Scandinavian ice caps. The same meridionality can explain the regional drying out of lakes and their origin in other, neighbouring regions (e.g. Kutzbach and Street-Perrot, 1985).

Global climate can thus be considered as a process of regional atmospheric warming and cooling and, related to this, of regional humidity variation. Both on fine and broad temporal scales zonal shifts occur near the equator and poles and of meandering, interdigitating outflows in between, along which cyclones and anticyclones travel as short-lived eddies. The resulting, highly dynamic spatial patterning of climatic conditions accounts for relatively sharply, though temporary defined regions. Biogeographical processes, both present and past, should be viewed against this background of regionally and globally climatic patterns, shifting on various time scales.

The representativity of our time

The fact that similar temperature and humidity conditions to those of the present day occur in less than 10% of the Quaternary, raises the problem of representativity of present-day observations. In fact, present vegetation has no analogue in the last glacial, or in previous interglacials, neither in temperate Europe and America, nor in tropical vegetations of the New Guinean mountains (Walker and Flenley, 1979). The present-day upper-mountain flora consists of mobile species able to occur for long glacial periods either in small, isolated pockets, or as components in the lower mountain forest. Present-day floral and faunal compositions were formed only after the end of the last glacial; they did not exist before, neither during the glacials, nor in previous interglacials (e.g. Azzaroli *et al.*, 1988; Birks, 1986; Huntley and Birks, 1983). Thus, certain species may form stable combinations through time, whereas others migrate independently (e.g. Davis, 1981*a*; Huntley and Birks, 1983; Wright, 1977). This telescoping of species migrating into and from other communities shows our limited knowledge of the compositional dynamics of communities that usually are assumed to be stable and closed, co-adapted species complexes. But can present-day communities of, for example, the mountain forest be estimated to be partly saturated to allow all the upper-mountain species to be taken up during the next glacial? How far do we understand communities analytically in terms of niche saturation? How far has niche theory already been substantiated for far-reaching generalizations? Do high spatio-temporal covariations exclusively indicate the species' ecological inter-dependences, or do they indicate ecological parameter similarity of the species concerned?

Similarly, when viewed at broad spatial scales, species can form a variety of different combinations. This means that a species' presence in one community does not predict whether it will occur in another community elsewhere or not; the species combination in which it occurs locally is not representative for all potential combinations over its range. Whenever detailed and long-term information on the spatial dynamics of many species is available, we observe continuous spatial adaptations to ever-changing conditions, each rate of adaptation being determined by a species' particular ecological properties and requirements. Present-day distribution patterns, in so far as they reflect climatic conditions, are not representative for longer periods of whatever ecologically relevant scale.

Conclusions

Biogeographical distributions of taxa at any level are characteristically broad-scale phenomena, suggesting that they can be explained in terms of variables that also vary on broad spatial scales. Usually, patterns at these scales are not repeated locally or at range margins; the explanatory factors found depend on the scale and locality chosen. Different processes, though occurring in the same species, such as its metabolism or dispersal rate, also depend on the same or on different factors, and operate on different scales as well. Moreover, for phenomena such as the species' geographical intensity distribution or its seasonality, several years are required for constructing reliable frequency distributions, making their characteristics depend on conditions prevalent in the period chosen. This dependence of results on chosen spatio-temporal scales also applies to species combinations. Thus species can covary in space at time scales with 1000 year intervals, but for many species this covariation breaks down for shorter intervals.

When effects of factors operating on a broad scale cannot be recognized on finer scales, for example, they should not be left unstudied. For example, when local differences in ground-water level explain the presence or absence of certain species, we may easily neglect effects of temperature when this is uniform among the sites. Had we compared observations from different parts of the geographical range, temperature variation would have been large enough to allow for the species' abundance variation on that scale, and intersite variation ascribed to variation in ground-water level would be considered as noise. Yet, the individuals are subjected to them both; they weight the effects of these variables according to their scales of operation as well as to the species' sensitivity regarding specific processes. Conversely, from the species' viewpoint, mutual adjustments are unlikely to arise. For example, competition, when it occurs, will usually be non-specific, specific inter-dependences being hazardous. Also, the risk that parasites run will, among other things, be determined by the temporal frequency distribu-tion of spatial overlap with its host's range, and hence by the similarity of their physiologies, reproduction rates, mobility, etc. This may explain why certain parts of the range of monophagous parasites, when they occur, are left unoccupied.

Regarding their genetic and ecological build-up, species vary on many spatio-temporal scales, thus expressing their sensitivity to factors operat-ing on different scales. They cannot be understood from observations on one single scale only, nor from their response to isolated factors. Species responses take many forms, species being physiologically rigid or flex-ible, and genetically fixed or adaptable. Their populations can fluctuate wildly or be stable, existing as a single or several ecologically distinct morphs, their geographical ranges can vary clinally or discontinuously, and their individuals can be sedentary or mobile and dispersive. All this together leads to variation in their intensity of occurrence on various spatial scales within the range and to changes in topographical character-istics of the range as such. The rate, direction, and inertia of all these processes depend on many environmental factors operating on several scales and varying either continuously or discontinuously in space and time. Biogeographical analysis should therefore allow for this multitude of factors and processes, weighted according to the species' sensitivity to each of them separately, in combination, and according to their scale of variation.

13

The future

Two lines of future research seem to be most fruitful: (1) the analysis of the eco-physiological basis of species ranges, species nests, etc., and (2) the analysis of population genetic, demographic, and spatial processes concerning the dynamics of species ranges. These lines together define a species' risk function as to its survival probability. Combining them can thus help us understand present-day biogeographical patterns and the processes operative within them.

Recent data-processing techniques, applied to large data sets, may allow us to understand such processes and can reveal and test their generality. But such sets can be biased for four reasons. (1) They may favour the dynamically most spectacular species such as plant and insect pests, or those living in the most changeable environments. The ranges of genetically or phenetically variable species can be more static than those of less variable species. (2) Large data sets often concern the more conspicuous taxa, such as butterflies, birds, plants, or beetles, whose spatial distribution and dynamics may differ from that of soil-dwelling or aquatic taxa. (3) Large data sets are mostly restricted to temperate regions, that is regions once covered with ice or lying close to its margin and which are colonized particularly by more dynamic species, resulting in relatively low degrees of endemism. Present populations north of 40°N are more affected by climatic fluctuations than those at lower latitudes. These regions are, moreover, comparatively flat. (4) Because temperate biotas contain a relatively high proportion of widely dispersed species and relatively few endemics, historical factors may be underestimated.

By discussing large data sets and giving this discussion an ecological impetus, I have emphasized the biology of biogeographical processes more than is usually done in discussions of biogeographical classification schemes, or of the geographical distribution of communities, biomes, formations or zonations. In my view, it is, to put it loosely, the species'

216

problem of how to maintain itself in a spatially heterogeneous and ever-changing environment that has to be explained. Biogeography therefore concerns, in the first instance, processes of species survival, and secondly how systems in a state of dynamic equilibrium evolve.

Genetical differentiation resulting in morphological, physiological and ecological variation within species ranges offers partial answers to these problems. But it has also its limitations, resulting in the spatially dome-shaped intensity distributions, in variation in habitat occupancy over the range, and in spatial dynamics of and within ranges. Ultimately, a species' genetic constitution determines its eco-physiological constitution, but not its ecological functioning resulting in its range characteristics, such as its location, structure, extension, and dynamics. These characteristics have to be analysed in their own right. Without these spatial and temporal aspects, species are still treated as non-dimensional entities, as in Hutchinson's (1978) concept of the ecological niche and in Whittaker's (1967) response curves according to gradient analysis. However, the axes spanning a niche hypervolume and representing environmental variables are oblique and have different weights, which both vary spatio-temporally. This makes the location of the various species within this hypervolume difficult to predict. Moreover, when, for example, climatic conditions change, all species move individualistically relative to each other within the hypervolume, or, in some parts of this volume, putting constraints on this mobility.

This complexity, of course, does not alter the basic idea of the Hutchinsonian niche, but it makes it extremely difficult to apply. Conversely, addition of spatio-temporal risk functions to characterize the degree to which a species' abundance distribution matches environmental variability, makes it feasible to formulate hypotheses as to causation and dynamic structure of species ranges, thus taking biogeography from historical explanation into the realm of ecological and genetical experimentation. In fact, laboratory or field experimentation is often the only way to determine the part history plays in biogeographical explanation. A species' geographical characteristics can best be studied from both its present-day ecology and dispersal capacity in relation to its fluctuating environment.

Apart from analysing phylogenetic relationships, biogeography is also concerned with evolution theory regarding range locations of related species relative to each other, isolation effects, or the existence of clines, emphasizing allopatric, sympatric, or some other kind of speciation. But when species ranges are ecological optimum surfaces, whose mean location and variance are partly determined through genetic fixation of ecological properties, geographical ranges can also be looked at from another evolutionary viewpoint. The recently discovered genetic structure and dynamics of species ranges is relevant here. However, present-day location patterns of species ranges hardly, if at all, show which species have probably split into one or more daughter species, which have migrated and which have died out, but the actual processes of splitting, migration, or extinction and their dependence on environmental fluctuation remain obscure. In future, biogeography and ecology should no longer be considered to be two distinct biological disciplines, working on different spatio-temporal scales with different factors operating and different processes occurring. Their subject is the same and their distinction artificial. Biogeographers should accept the idea that species ranges are dynamic structures, rather than unstructured, monolithic static entities. Ecologists, on their part, should adjust to the idea that spatially and temporally broad-scale processes also occur, being of paramount importance for the values that both synecological and autecological parameters take locally and define the course of population dynamic processes. A more geographical outlook provides the context for a species' fate; increases or decreases in local densities can reflect geographical shifts in range location, rather than indicate the onset of outbreak proportions, or local extinction only. At all scales, species ranges should not be looked upon as static and homogeneous entities, forming stable ecological communities of coadapted species, but as flexible and highly dynamic and individualistic structures in a continuous state of flux.

References

Abele, L.G., K.L. Heck, D.S. Simberloff and G.J. Vermey (1981). Biogeography of crab claw size: assumptions and a null hypothesis. *Systematic Zoology*, **30**, 406–24.

Alkon, P.U. and D. Saltz (1988). Foraging time and the northern range limits of Indian crested porcupines (*Hystrix indica* Kerr). *Journal of Biogeography*, **15**, 403–8.

Allard, R.W., R.D. Miller and A.L. Kahler (1978). The relationship between degree of environmental heterogeneity and genetic polymorphism. In *Structure and functioning of plant populations*, eds. A.H.J. Freysen and J.W. Woldendorp, pp. 49–73. North Holland Publishing Company, Amsterdam.

Allen, J.A. (1877). The influence of physical conditions in the genesis of species. *Radical Review*, **1**, 108–40.

Amadon, D. (1949). The seventy-five percent rule for species. *Condor*, **51**, 250–8.

Ammerman, A.J. and L.L. Cavalli-Sforza (1984). *The neolithic transition and the genetics of populations in Europe*. Princeton University Press, Princeton.

Anderson, T.W. (1974). The chestnut pollen decline as a time horizon in lake sediments in eastern North America. *Canadian Journal of Earth Sciences*, **11**, 678–85.

Andrewartha, H.G. and L.C. Birch (1954). *The distribution and abundance of animals*. Chicago University Press, Chicago.

— (1984). *The ecological web*. Chicago University Press, Chicago.

Antonovics, J. (1978). The population genetics of mixtures. In *Plant relations in pastures*, ed. J.R. Wilson, pp. 233–52, East Melbourne.

Ås, S. (1984). To fly or not to fly? Colonization of Baltic islands by winged and wingless carabid beetles. *Journal of Biogeography*, **11**, 413–26.

Azzaroli, A., C. de Giuli, G. Ficcarelli and D. Torre (1988). Late Pleistocene to early Mid-Pleistocene mammals in Eurasia: faunal succession and dispersal events. *Palaeogeography, Palaeoclimatology, Palaeoecology*, **66**, 77–100.

Babbel, G.R. and R.K. Selander (1974). Genetic variability in edaphically restricted and widespread plant species. *Evolution*, **28**, 619–30.

Backhuys, W. (1968). Der Elevations-effect bei einigen Alpenpflanzen der Schweiz. *Blumea*, **16**, 273–320.

Bailey, J.W. and E.W. Sinnott (1916). The climatic distribution of certain types of angiosperm leaves. *American Journal of Botany*, **3**, 24–39.

Baker, H.G. (1959). The contribution of autecological and genecological studies to our knowledge of the past migrations of plants. *American Naturalist*, **93**, 255–72.

Bartlein, P.J., I.C. Prentice and T. Webb (1986). Climatic response surfaces for some

219

eastern North American pollen types. *Journal of Biogeography*, **13**, 35–57.

Barton, N.H., J.S. Jones and J. Mallet (1988). No barriers to speciation. *Nature*, **336**, 13–4.

Beauchamp, R.S.A., and P. Ullyott (1932). Competitive relationships between certain species of fresh-water triclades. *Journal of Ecology*, **20**, 200–8.

Bennett, M.D. (1972). Nuclear DNA content and minimum generation time in herbaceous plants. *Proceedings of the Royal Society of London*, B **181**, 109–35.

— (1976). DNA amount, latitude, and crop plant distribution. *Environmental Experimental Botany*, **16**, 93–108.

Bergmann, C. (1847). Ueber die Verhältnisse der Warmeökologie der Thiere zu ihrer Grosse. *Göttinger Studien*, **1**, 595–708.

Bergthorsson, P. (1985). Sensitivity of Icelandic agriculture to climatic variations. *Climatic Change*, **7**, 111–27.

Beven, S., E.F. Connor and K. Beven (1984). Avian biogeography in the Amazon basin and the biological model of diversification. *Journal of Biogeography*, **11**, 383–99.

Billings, W.D. (1952). The environmental complex in relation to plant growth and distribution. *Quarterly Review of Biology*, **27**, 251–65.

Birch, L.C., Th. Dobzhansky, P.O. Eliott and R.C. Lewontin (1963). Relative fitness of geographic races of *Drosophila serrata*. *Evolution*, **17**, 72–83.

Birks, H.J.B. (1976). The distribution of European pteridophytes: a numerical analysis. *New Phytologist*, **77**, 257–87.

— (1986). Late-Quaternary biotic changes in terrestrial and lacustrine environments, with particular reference to North-West Europe. In *Handbook of palaeoecology and palaeohydrology*, ed. B.E. Berglund, pp. 3–65. Wiley, New York.

— (1987). Recent methodological developments in quantitative descriptive biology. *Annales Zoologici Fennici*, **24**, 165–77.

— J. Deacon and S. Peglar (1975). Pollen maps for the British Isles 5000 years ago. *Proceedings of the Royal Society of London*, B **189**, 87–105.

Black, F.L. (1966). Measles endemicity in insular populations: critical community size and its evolutionary implication. *Journal of Theoretical Biology*, **11**, 207–11.

Blem, C.R. (1973). Geographic variation in the bioenergetics of the house sparrow. *Ornithological Monographs*, **14**, 96–121.

— (1980). The energetics of migration. In *Animal migration, orientation, and navigation*, ed. S.A. Gauthreaux, pp. 175–224. Academic Press, New York.

Bock, C.E. (1984). Geographical correlates of abundance vs. rarity in some North American winter land birds. *Auk*, **101**, 266–73.

— and R.E. Ricklefs (1983). Range size and local abundance of some North American songbirds: a positive correlation. *American Naturalist*, **122**, 295–9.

Borchert, J.R. (1950). The climate of the central North American grassland. *Annals of the Association of American Geographers*, **40**, 1–39.

— (1971). The dust bowl in the 1970s. *Annals of the Association of American Geographers*, **61**, 1–22.

Bowers, M.A. (1988). Relationships between local distribution and geographic range of desert heteromyid rodents. *Oikos*, **53**, 303–8.

Bradka, J. (1966). Natural seasons on the Northern Hemisphere. *Geofysikalni Sbornik*, **14**, 597–648.

Braun-Blanquet, J. (1932). *Pflanzensoziologie*. English edition, 1972. McGraw-Hill, New York.

Briggs, J.C. (1974). The operation of zoogeographic barriers. *Systematic Zoology*, **23**, 248–56.

— (1984). *Centres of origin in biogeography*. Biogeography Study Group, Leeds.

Broekhuizen, S., ed. (1969). *Agro-ecological atlas of cereal growing in Europe*. Vol. 2. Elsevier, Amsterdam.

Brown, J.H. (1984). On the relationship between abundance and distribution of species. *American Naturalist*, **124**, 255–79.

Brown, W.L. (1957). Centrifugal speciation. *Quarterly Review of Biology*, **32**, 247–77.

Brugam, R.B. (1978). Pollen indicators of land-use changes in southern Connecticut. *Quaternary Research*, **9**, 349–62.

Bryson, R.A. (1966). Air masses, streamlines, and the boreal forest. *Geographical Bulletin*, **8**, 228–69.

— and F.K. Hare (1974). The climates of North America. In *Climates of North America*, eds. R.A. Bryson and F.K. Hare, pp. 1–47. Elsevier, Amsterdam.

— W.N. Irving and J.A. Larsen (1965). Radiocarbon and soil evidence of former forest in the southern Canadian tundra. *Science*, **147**, 46–8.

Burrough, P.A. (1981). Fractal dimensions of landscapes and other environmental data. *Nature*, **294**, 240–2.

— (1983*a*). Multiscale sources of spatial variation in soil. I. The application of fractal concepts to tested levels of soil variation. *Journal of Soil Science*, **34**, 577–97.

— (1983*b*). Multiscale sources of spatial variation in soil. II. A non-Brownian fractal model and its application in soil survey. *Journal of Soil Science*, **34**, 599–620.

— (1986). *Principles of geographical information systems for land resource assessment*. Oxford University Press, Oxford.

Bush, M.B. (1988). The use of multivariate analysis and modern analogue sites as an aid to the interpretation of data from fossil mollusc assemblages. *Journal of Biogeography*, **15**, 849–61.

Cain, S.A. (1944). *Foundations of plant geography*. Harper and Brothers, New York.

Calder, W.A. (1984). *Size, function and life history*. Harvard University Press, Cambridge, Mass.

Carlquist, S. (1974). *Island biology*. Columbia University Press, New York.

Carne, P.B. (1965). Distribution of the *Eucalypt*-defoliating sawfly *Perga affinis affinis* (*Hemiptera*). *Australian Journal of Zoology*, **13**, 593–612.

Carter, R.N. and S.D. Prince (1981). Epidemic models used to explain biogeographical distribution limits. *Nature*, **293**, 644–5.

— and — (1985). The geographical distribution of prickly lettuce (*Lactuca serriola*). I. A general survey of its habitats and performance in Britain. *Journal of Ecology*, **73**, 27–38.

Caughley, G. (1963). Dispersal rates of several ungulates introduced in New Zealand. *Nature*, **200**, 280–1.

— D. Grice, R. Barker and B. Brown (1988). The edge of the range. *Journal of Animal Ecology*, **57**, 771–85.

— J. Short, G.C. Grigg and H. Nix (1987). Kangaroos and climate: an analysis of distribution. *Journal of Animal Ecology*, **56**, 751–61.

Cavagnaro, J.B. (1988). Distribution of C3 and C4 grasses at different altitudes in a temperate arid region of Argentina. *Oecologia*, **76**, 273–7.

Chazdon, R.L. (1978). Ecological aspects of the distribution of C4 grasses in selected habitats of Costa Rica. *Biotropica*, **10**, 265–9.

Cheetham, A.H. and J.E. Hazel (1969). Binary (presence–absence) similarity coefficients. *Journal of Paleontology*, **43**, 1130–6.

Clark, W.C., D.D. Jones and C.S. Holling (1978). Patches, movements, and population dynamics in ecological systems: a terrestrial perspective. In *Spatial pattern in plankton communities*, ed. J.H. Steele. New York.

Clarke, L. R. (1953). The ecology of *Chrysomela gemelata* Rossi and *C. hyperici* Forst., and their effect on St John's wort in the Bright district, Victoria. *Australian Journal of Zoology*, **1**, 1–69.

Clayton, W.D. (1976). The chorology of African mountain grasses. *Kew Bulletin*, **31**, 273–88.

— and G. Panigrahi (1974). Computer-aided chorology of Indian grasses. *Kew Bulletin*, **29**, 669–86.

Cliff, A.D., P. Haggett, J.K. Ord and G.R. Versey (1981). *Spatial diffusion*. Cambridge University Press, Cambridge.

Clifford, H.T. and W. Stephenson (1975). *An introduction to numerical classification*. Academic Press, New York.

CLIMAP (1976). Climate mapping and prediction. *Science*, **191**, 1131–7.

COHMAP (1988). Climatic changes of the last 18,000 years: observations and simulations. *Science*, **241**, 1043–52.

Connor, E.F. and D. McCoy (1979). The statistics and biology of the species–area relationship. *American Naturalist*, **113**, 791–833.

— and D. Simberloff (1978). Species number and compositional similarity of the Galapagos flora and fauna. *Ecological Monographs*, **48**, 219–48.

— and — (1979). The assembly of species communities: chance or competition? *Ecology*, **60**, 1132–40.

Cook, R.E. (1967). Variation in species density of North American birds. *Systematic Zoology*, **16**, 63–84.

Cook, W.C. (1925). The distribution of the alfalfa weevil (*Phytonomus posticus* Gyll.). A study in physical ecology. *Journal of Agricultural Research*, **30**, 479–91.

— (1929). A bioclimatic zonation for studying the economic distribution of injurious insects. *Ecology*, **10**, 282–93.

Coombs, C.H. (1964). *A theory of data*. Wiley, New York.

Coope, G.R. (1975). Mid-Weichselian climatic changes in Western Europe, re-interpreted from coleopteran assemblages. In *Quaternary studies*, eds. R.P. Suggate and M.M. Cresswell, pp. 101–8. Wellington.

Cooper, J.P. (1970). Environmental physiology. In *Genetic resources in plants – their exploration and conservation*, eds. O.H. Frankel and E. Bennett, pp. 131–42. Blackwell, Oxford.

Cormack, R.M. (1971). A review of classification. *Journal of the Royal Statistical Society* (A), **134**, 321–67.

Coupland, R.T. (1958). The effects of fluctuations in weather upon the grasslands of the Great Plains. *Botanical Review*, **24**, 273–317.

— (1959). Effects of changes in weather conditions upon grasslands in the northern Great Plains. In *Grasslands*, ed. H.B. Sprague.

— (1974). Fluctuations in North American grassland vegetation. In *Vegetation dynamics*, ed. R. Knapp, pp. 233–41. Junk, The Hague.

Crawley, M.J. (1986). What makes communities stable? In *Colonization, succession and stability*, eds. M.J. Crawley, P.J. Edwards and A.J. Gray, pp. 429–53. Blackwell, Oxford.

Currey, J.D. and A.J. Cain (1968). Studies on *Cepaea*. IV. Climate and selection of banding morphs in *Cepaea* from the climatic optimum to the present day. *Philosophical Transactions of the Royal Society of London*, B.**253**, 483–98.

Currie, D.J. and V. Paquin (1987). Large-scale biogeographical patterns of species richness of trees. *Nature*, **329**, 326–7.

Curtis, J.T. (1959). *The vegetation of Wisconsin*. University of Wisconsin Press, Madison.

Cushing, D.H. (1982). *Climate and fisheries*. Academic Press, London.
— and R.R. Dickson (1976). The biological response in the sea to climatic changes. *Advances in Marine Biology*, **14**, 1–122.
Dahl, E. (1951). On the relation between summer temperature and the distribution of alpine plants in the lowlands of Fennoscandia. *Oikos*, **3**, 22–52.
Dale, M.R.T. (1988). The spacing and intermingling of species boundaries on an environmental gradient. *Oikos*, **53**, 351–6.
Danilevskii, A.S. (1965). *Photoperiodism and seasonal development in insects*. Oliver and Boyd, Edinburgh.
Darlington, P.J. (1957). *Zoogeography*. Wiley, New York.
Davis, M.B. (1981*a*). Quaternary history and the stability of forest communities. In *Forest succession*, eds. D.C. West, H.H. Shugart and D.B. Botkin, pp. 132–53. Springer, New York.
— (1981*b*). *Outbreaks of forest pathogens in Quaternary history*. IV. *International Palynological Conference, Lucknow (1976–1977)*, **3**, 216–28.
— (1986). Climatic instability, time lags, and community disequilibrium. In *Community ecology*, eds. J. Diamond and T.J. Case, pp. 269–84. Harper and Row, New York.
Davis, R. and C. Dunford (1987). An example of temporary colonization of montane islands by small, nonflying mammals in the American Southwest. *American Naturalist*, **129**, 398–406.
—, — and M.V. Lomolino (1988). Montane mammals of the American Southwest: the possible influence of post-Pleistocene colonization. *Journal of Biogeography*, **15**, 841–48.
De Lattin, G. (1957). Die Ausbreitungszentren der holarktischen Landtierwelt. *Verhandlungen des deutschen zoologischen Gesellschafts*, **1956**, 380–410.
Delcourt, H.R., P.A. Delcourt and T. Webb (1983). Dynamic plant ecology: the spectrum of vegetational change in space and time. *Quaternary Science Review*, **1**, 153–75.
Den Boer, P.J. (1968). Spreading of risk and stabilization of animal numbers. *Acta Biotheoretica*, **18**, 165–94.
Diamond, J.M. (1975). Assembly of species communities. In *Ecology and evolution of communities*, eds. M.L. Cody and J.M. Diamond, pp. 342–444. Harvard University Press, Cambridge, Mass.
— and M.E. Gilpin (1982). Examination of the 'null' model of Connor and Simberloff for species co-occurrences on islands. *Oecologia*, **52**, 64–74.
Dobson, A.P. and R.M. May (1986). Patterns of invasions by pathogens and parasites. In *Ecology of biological invasions of North America and Hawaii*, eds. H.A. Mooney and J.A. Drake, pp. 58–76. Springer, New York.
Dobzhansky, T. (1950). Evolution in the tropics. *American Scientist*, **38**, 208–21.
Dodd, A.P. (1959). The biological control of the prickly pear in Australia. In *Biogeography and ecology in Australia*, eds. A. Keast, R.L. Crocker and C.S. Christian, pp. 565–77. *Monographiae Biologicae*, Vol. 8. Junk, The Hague.
Drude, O. (1876). *Die Anwendung physiologischer Gesetze zur Erklärung der Vegetationslinien*. Göttingen.
Durazzi, J.T. and F.G. Stehli (1972). Average generic age, the planetary temperature gradient, and pole location. *Systematic Zoology*, **21**, 384–9.
Edwards, G. and D.A. Walker (1983). *C3, C4: mechanisms, and cellular and environmental regulation, of photosynthesis*. Oxford University Press, Oxford.
Egerton, F.N. (1973). Changing concepts of the balance of nature. *Quarterly Review of Biology*, **48**, 322–50.

Ehrendorfer, F. (1951). Rassengliederung, Variabilitätszentren und geographische Merkmalsprogression als Audruck der raumzeitlichen Entfaltung des Formenkreises *Galium incanum* s.s. *Oestereichisches botanisches Zeitschrift*, **100**, 583–92.

— (1958). Die geographische und ökologische Entfaltung des europäisch-alpinen Polyploidkomplexes *Galium anisophyllum* Vill. seit Beginn des Quartärs. *Uppsala Universitäts Årsskrift*, **1958** (6), 176–81.

Eidmann, H. (1949). Verbreitung und Schadgebiet des Tannenwicklers *Cacoecia murinana* Hb. (Lep., Tortricidae). *Anzeiger fur Schädlingskunde*, **22**, 103–7.

Ekman, S. (1953). *Zoogeography of the sea.* Sidgwick and Jackson, London.

Elton, C.S. (1958). *The ecology of invasions by animals and plants.* Methuen, London.

Emlen, J.M. (1973). *Ecology: an evolutionary approach.* Addison Wesley, Reading, Mass.

Endler, J.A. (1982). Problems in distinguishing historical from ecological factors in biogeography. *American Zoologist*, **22**, 441–52.

Enright, J.T. (1976). Climate and population regulation, the biogeographer's dilemma. *Oecologia*, **24**, 295–310.

Erkamo, V. (1956). Untersuchungen ueber die pflanzenbiologischen und einige andere Folgeerscheinungen der neuzeitlichen Klimaschwankung in Finnland. *Annales Botanici Societas Zoologicae Botanicae Fennicae 'Vanamo'*, **28**, 1–290.

Ezcurra, E., E.H. Rapoport and C.R. Marino (1978). The geographical distribution of insect pests. *Journal of Biogeography*, **5**, 149–57.

Fernald, M.L. (1926). The antiquity and dispersal of vascular plants. *Quarterly Review of Biology*, **1**, 212–45.

Fernandes, G.W. and P.W. Price (1988). Biogeographical gradients in galling species richness. *Oecologia*, **76**, 161–7.

Fischer, A.G. (1960). Latitudinal variations in organic diversity. *Evolution*, **14**, 64–81.

Fisher, R.A., A.S. Corbet and C.B. Williams (1943). The relation between the number of species and the number of individuals in a random sample of an animal population. *Journal of Animal Ecology*, **12**, 42–58.

Fleischer, R.C. and R.F. Johnston (1982). Natural selection on body size and proportions in house sparrows. *Nature*, **298**, 747–9.

Flenley, J.R. and K. Richards (1982). *The Krakatoa centenary expedition. Final report.* University of Hull, Dept. of Geography, Miscellaneous Series, 25.

Ford, M.J. (1982). *The changing climate.* Allen and Unwin, London.

Franz, J.M. (1964). Dispersion and natural-enemy action. *Proceedings of the Association of Applied Biology*, **53**, 510–5.

Gaston, K.J. and J.H. Lawton (1988a). Patterns in the distribution and abundance of insect populations. *Nature*, **321**, 709–12.

— and — (1988b). Patterns in body size, population dynamics, and regional distribution of bracken herbivores. *American Naturalist*, **132**, 662–80.

Gause, G.F. (1930). Studies on the ecology of the Orthoptera. *Ecology*, **11**, 307–25.

— (1932). Ecology of populations. *Quarterly Review of Biology*, **7**, 27–46.

Gauslaa, Y. (1984). Heat resistance and energy budget in different Scandinavian plants. *Holarctic Ecology*, **7**, 1–78.

Ghilarov, M. (1959). *Die Gesetzmässigkeiten der zonalen Verbreitung schädlicher Bodeninsekten im europäischen Teile der UdSSR. 4. Internationaler Pflanzenschutzkongress, Hamburg 1957*, **1**, 831–6.

Glazier, D.S. (1980). Ecological shifts and the evaluation of geographically restricted species of North American *Peromyscus* (mice). *Journal of Biogeography*, **7**, 63–83.

— (1988). Temporal variability of abundance and the distribution of species. *Oikos*, **47**, 309–14.

Gleason, H.A. (1939). The individualistic concept of the plant association. *American Midland Naturalist*, **21**, 92–110.

Gloger, C.L. (1833). *Das Abändern der Vogel durch Einfluss des Klimas*. Breslau.

Gloyne, R.W. (1972). The 'growing season' at Eskdalemuir observatory, Dumfriesshire. *Meteorological Magazine*, **102**, 174–8.

Good, R.A. (1931). A theory of plant geography. *New Phytologist*, **30**, 149–71.

— (1974). *The geography of flowering plants*, 4th ed. Longman, London.

Goodall, D.W. (1954). Objective methods for the classification of vegetation. III. An essay in the use of factor analysis. *Australian Journal of Botany*, **2**, 304–24.

— (1964). A probabilistic similarity index. *Nature*, **203**, 1098.

— (1966). A new similarity index based on probability. *Biometrics*, **22**, 883–907.

— (1986). Classification and ordination: their nature and role in taxonomy and community studies. *Coenosis*, **1**, 3–9.

Gordon, A.D. (1981). *Classification*. Chapman and Hall, London.

Gottschalk, W. (1976). *Die Bedeutung der Polyploidie für die Evolution der Pflanzen*, Fischer Verlag, Stuttgart.

Gower, J.C. (1967). Multivariate analysis and multidimensional geometry. *The Statistician*, **17**, 13–28.

Graham, R.W. (1986). Response of mammalian communities to environmental changes during the late Quaternary. In *Community ecology*, eds. J.D. Diamond and T.J. Case, pp. 300–13. Harper and Row, New York.

Grant, V. (1959). *Natural history of the Phlox family*. Nijhoff, The Hague.

Greenbank, D.O. (1956). The role of climate and dispersal in the initiation of outbreaks of the spruce budworm in New Brunswick. I. The role of climate. *Canadian Journal of Zoology*, **34**, 453–76.

— (1957). The role of climate and dispersal in the initiation of outbreaks of the spruce budworm in New Brunswick. II. The role of dispersal. *Canadian Journal of Zoology*, **35**, 385–403.

— G.W. Schaefer and R.C. Rainey (1980). Spruce budworm (Lepidoptera: Tortricidae) moth flight and dispersal: new understanding from canopy observations, radar, and aircraft. *Memoirs of the Entomological Society of Canada*, **110**, 1–49.

Grime, J.P. and M.A. Mowforth (1982). Variation in genome size – an ecological interpretation. *Nature*, **299**, 151–3.

Guthrie, R.D. (1984). Mosaics, allelochemics, and nutrients: an ecological theory of Late Pleistocene megafaunal extinctions. In *Quaternary extinctions*, eds. P.S. Martin and R.G. Klein, pp. 259–98. Arizona University Press, Tuscon.

Haeck, J. and R. Hengeveld (1977). Distribution ecology of carabid beetles. Institutes of the Royal Netherlands Academy of Arts and Sciences, Progress Report, 176, Institute of Ecological Research. *Verhandelingen der Koninklijke Nederlandse Akademie van Wetenschappen, Afd. Natuurkunde, 2e Reeks*, **69**, 24–7.

— and — (1980). Changes in the occurrences of Dutch plant species in relation to geographical range. *Biological Conservation*, **19**, 189–97.

Haeupler, H. (1974). Statistische Auswertung von Punktrasterkarten der Gefässpflanzenflora süd-Niedersachsens. *Scripta Geobotanica*, **8**, 1–141.

Haffer, J. (1969). Speciation in Amazonian forest birds. *Science*, **165**, 131–7.

Hagmeier, E.M. (1966). A numerical analysis of the distributional patterns of North America. II. Re-evaluation of the provinces. *Systematic Zoology*, **15**, 279–99.

— and C.D. Stults (1964). A numerical analysis of the distributional patterns of North American mammals. *Systematic Zoology*, **13**, 125–55.

226   *References*

Hanski, I. (1982). Dynamics of regional distribution: the core and satellite species hypothesis. *Oikos*, **38**, 210–21.

Harlan, J.R. and D. Zohari (1966). Distribution of wild wheats and barley. *Science*, **153**, 1074–80.

Harper, C.W. (1978). Groupings by locality in community ecology and paleoecology: tests of significance. *Lethaia*, **11**, 251–7.

Harris, P., D. Peschken and J. Milroy (1969). The status of biological control of the weed *Hypericum perforatum* in British Columbia. *Canadian Entomologist*, **101**, 1–15.

Hartley, W. (1970). Climate and crop distribution. In *Genetic resources in plants – their exploration and conservation*, eds. O.H. Frankel and E. Bennett, pp. 143–53. Blackwell, Oxford.

Hattersley, P.W. (1983). The distribution of C3 and C4 grasses in Australia in relation to climate. *Oecologia*, **57**, 113–28.

Heads, P.A. and J.H. Lawton (1983). Studies on the natural enemy complex of the holly leaf-miner: the effects of scale on the detection of aggregative responses and the implications for biological control. *Oikos*, **40**, 267–76.

Hecht, A.D. and B. Agon (1972). Diversity and age relationships in recent and Miocene bivalves. *Systematic Zoology*, **21**, 308–12.

Heggberget, T.M. (1987). Number and proportion of southern bird species in Norway in relation to latitude, spring temperature and respiration coefficient. *Holarctic Ecology*, **10**, 81–9.

Henderson, R.A. and M.L. Heron (1977). A probabilistic method of paleobiogeographic analysis. *Lethaia*, **10**, 1–15.

Henderson-Sellers, A. (1986). Cloud changes in a warmer Europe. *Climatic Change*, **8**, 25–52.

Hengeveld, R. (1985a). On the explanation of the elevation effect by a dynamic interpretation of species distribution along altitudinal gradients. *Blumea*, **30**, 353–61.

— (1985b). Dynamics of Dutch ground beetle species during the twentieth century. *Journal of Biogeography*, **12**, 389–411.

— (1985c). Methodology of explaining differences in dietary composition of carabid beetles by competition. *Oikos*, **45**, 37–49.

— (1987a). Scales of variation: their distinction and ecological importance. *Acta Zoologica Fennica*, **24**, 195–202.

— (1987b). Theories on biological invasions. *Proceedings of the Koninklijke Nederlandse Akademie van Wetenschappen*, C **90**, 45–9.

— (1988a). Mayr's ecological species criterion. *Systematic Zoology*, **37**, 47–55.

— (1988b). Mechanisms of biological invasions. *Journal of Biogeography*, **15**, 819–28.

— (1989a). *Dynamics of biological invasions*. Chapman and Hall, London.

— (1989b). Caught in an ecological web. *Oikos*, **54**, 15–22.

—, H.B. Becker and H. van Biezen (1982). Aspekten van statistisch gedrag van diversiteitsmaten. *Vakblad voor Biologen*, **62**, 230–4.

— and J. Haeck (1981). The distribution of abundance. II. Models and implications. *Proceedings of the Koninklijke Nederlandse Akademie van Wetenschappen*, C**84**, 257–84.

— and — (1982). The distribution of abundance. I. Measurements. *Journal of Biogeography*, **9**, 303–16.

— and P. Hogeweg (1979). Cluster analysis of the distribution patterns of Dutch carabid species (Col.). In *Multivariate methods in ecological work*, eds. L. Orloci,

C.R. Rao and W.M. Stiteler, pp. 65–86. International Co-operative Publishing House, Fairland, Maryland.

—, S.A.L.M. Kooijman and C. Taillie (1979). A spatial model explaining species-abundance curves. In *Statistical distributions in ecological work*, eds. J.K. Ord, G.P. Patil and C. Taillie, pp. 333–47. International Co-operative Publishing House, Fairland, Maryland.

— and A.J. Stam (1978). Models explaining Fisher's log-series abundance curve. *Proceedings of the Koninklijke Nederlandse Akademie van Wetenschappen*, C81, 415–27.

Higgins, L.G. and N.D. Riley (1970). *A field guide to the butterflies of Britain and Europe*. Collins, London.

Hill, M.O. (1979). *TWINSPAN – A FORTRAN program for arranging multivariate data in an ordered two-way table by classification of individuals and attributes*, Cornell University, Ithaca, N.Y.

Hill, R.S., J. Read and J.R. Busky (1988). The temperature-dependence of photosynthesis of some Australian temperature rainforest trees and its biogeographical significance. *Journal of Biogeography*, 15, 431–49.

Hintikka, V. (1963). Ueber das Grossklima einiger Pflanzenareale, durch zwei Klimakoordinatensystemen dargestellt. *Annales Botanici Societatis Zoologicae Botanicae Fennicae 'Vanamo'*, 34, 1–65.

Holloway, J.D. (1979). *A survey of the Lepidoptera, biogeography and ecology of New Caledonia*. Junk, The Hague.

— (1982). Mobile organisms in a geologically complex area: Lepidoptera in the Indo-Australian tropics. *Zoological Journal of the Linnean Society*, 76, 353–73.

— and N. Jardine (1968). Two approaches to zoogeography: a study based on the distributions of butterflies, birds and bats in the Indo-Australian area. *Proceedings of the Linnean Society of London*, 179, 153–88.

Holzner, W. (1978). Weed species and weed communities. In *Plant species and plant communities*, eds. E. van der Maarel and M.J.A. Werger, pp. 119–26. Junk, The Hague.

Horn, H.S. (1966). Measurement of 'overlap' in comparative ecological studies. *American Naturalist*, 100, 419–24.

Horn, L.H. and R.A. Bryson (1960). Harmonic analysis of the annual march of precipitation over the United States. *Annals of the Association of American Geographers*, 50, 157–71.

Howe, H.F. and J. Smallwood (1982). Ecology and seed dispersal. *Annual Review of Ecology and Systematics*, 13, 201–28.

Huffaker, C.B. and C.E. Kennett (1959). A ten-year study of vegetational changes associated with biological control of Klamath weed. *Journal of Range Management*, 12, 69–82.

Hughes, C.P. (1973). Analysis of pest faunal distributions. In *Implications of continental drift to the earth sciences*, eds. D.H. Tarling and S.K. Runcorn, pp. 221–30. London.

Hultén, E. (1937). *Outline of the history of arctic and boreal biota during the Quaternary period*. Cramer, Stockholm.

Huntley, B. and H.J.B. Birks (1983). *An atlas of past and present pollen maps for Europe: 0–13 000 years ago*. Cambridge University Press, Cambridge.

— and I.C. Prentice (1988). July temperatures in Europe from pollen data, 6000 years before present. *Science*, 241, 687–90.

Hutchins, L.W. (1947). The bases for temperature zonation in geographical distribution. *Ecological Monographs*, 17, 325–35.

Hutchinson, G.E. (1978). *An introduction to population ecology*. Yale University Press, New Haven.

Imbrie, J. and N.G. Kipp (1971). A new micropaleontological method for quantitative paleontology: application to a late Pleistocene Caribbean core. In *Late Cenozoic glacial ages*, ed. K.K. Turekian, pp. 71–181. Yale University Press, New Haven.

Iversen, J. (1944). *Viscum, Hedera* and *Ilex* as climate indicators. *Geologiska Föreningens i Stockholm forhandlingar*, **66**, 463–83.

Jaccard, P. (1908). Nouvelles recherches sur la distribution florale. *Bulletin de la Société Vaudoise de la Science Naturelle*, **44**, 223–70.

Jablonski, D. (1986). Background and mass extinctions: the alternation of macroevolutionary regimes. *Science*, **231**, 129–33.

Jackson, J.B.C. (1974). Biogeographic consequences of eurytopy and stenotopy among marine bivalves and their evolutionary significance. *American Naturalist*, **108**, 541–60.

James, F.C. (1970). Geographic size variation in birds and its relationship to climate. *Ecology*, **51**, 365–90.

James, J.C. (1983). Environmental component of morphological differentiation in birds. *Science*, **221**, 184–6.

Jardine, N. (1971). Patterns of differentiation between human populations. *Philosophical Transactions of the Royal Society*, B **263**, 1–33.

— and J.M. Edmonds (1974). The use of numerical methods to describe population differentiation. *New Phytologist*, **73**, 1259–77.

— and R. Sibson (1971). *Mathematical taxonomy*. Wiley, London.

Järvinen, O. and R.A. Vaisänen (1980). Quantitative biogeography of Finnish land birds as compared with regionality in other taxa. *Annales Zoologici Fennici*, **17**, 67–85.

Jeffree, E.P. (1969). Some long-term means from the phenological reports (1891–1948) of the Royal Meteorological Society. *Quarterly Journal of the Royal Meteorological Society*, **86**, 95–103.

Jerling, L. (1985). Population dynamics of *Plantago maritima* along a distributional gradient on a Baltic seashore meadow. *Vegetatio*, **61**, 155–61.

— (1988). Genetic differentiation in fitness related characters in *Plantago maritima* along a distributional gradient. *Oikos*, **53**, 341–50.

Johnston, R.F. and R.K. Selander (1971). Evolution in the house sparrow, II. Adaptive differentiation in North American populations. *Evolution*, **25**, 1–28.

Kaiser, G.W., L.P. Lefkovitch and H.F. Howden (1972). Faunal provinces in Canada as exemplified by mammals and birds: a mathematical consideration. *Canadian Journal of Zoology*, **50**, 1087–104.

Kareiva, P. (1986). Patchiness, dispersal, and species interactions: consequences for communities of herbivorous insects. In *Community ecology*, eds. J. Diamond and T.J. Case, pp. 192–206. Harper and Row, New York.

— and G. Odell (1987). Swarms of predators exhibit 'prey taxis' if individual predators use area-restricted search. *American Naturalist*, **130**, 233–70.

Kavanagh, K. and M. Kellman (1986). Performance of *Tsuga canadensis* (L.) Carr. at the centre and northern edge of its range: a comparison. *Journal of Biogeography*, **13**, 145–57.

Kempton, R.A. (1981). The stability of site ordinations in ecological surveys. In *The mathematical theory of the dynamics of biological populations*. II. eds. R.W. Hiorns and C. Cooke, pp. 217–30. Academic Press, London.

Kendeigh, S.C. (1944). Effect of air temperature on the rate of energy metabolism in the English sparrow. *Journal of Experimental Zoology*, **96**, 1–16.

— (1949). Effect of temperature and season on energy resources of the English sparrow. *Auk*, **66**, 113–27.

— (1969). Tolerance of cold and Bergmann's Rule. *Auk*, **86**, 13–25.

— (1976). Latitudinal trends in the metabolic adjustments of the house sparrow. *Ecology*, **57**, 509–19.

Kennedy, C.H. (1928). Evolutionary level in relation to geographical, seasonal and diurnal distribution of insects. *Ecology*, **9**, 367–79.

Kikkawa, J. and K. Pearse (1969). Geographical distribution of land birds in Australia – a numerical analysis. *Australian Journal of Zoology*, **17**, 821–40.

Klages, K.H.W. (1942). *Ecological crop geography*. MacMillan, New York.

Kless, J. (1961). Tiergeographische Elemente in der Käfer- und Wanzenfauna des Wutachgebietes und ihre oekologischen Anspruche. *Zeitschrift für Morphologie und Oekologie der Tiere*, **49**, 541–628.

Kooijman, S.A.L.M. (1989). The Von Bertalanffy growth rate as a function of physiological parameters. (In preparation.)

— and R. Hengeveld (1979). The description of a non-linear relationship between some carabid beetles and environmental factors. In *Contemporary quantitative ecology and related ecometrics*, eds. G.P. Patil and M.L. Rosenzweig, pp. 635–47. International Co-operative Publishing House, Fairland, Maryland.

Koskimies, J. and L. Lahti (1964). Cold-hardiness of the newly hatched young in relation to ecology and distribution in ten species of European ducks. *Auk*, **81**, 281–307.

Kozar, F. and A.N. David (1986). The unexpected northward migration of some species of insects in Central Europe and the climatic changes. *Anzeiger für Schädlingskunde, Pflanzenschutz und Umweltschutz*, **59**, 90–4.

Krebs, J.S. and R.G. Barry (1970). The arctic front and the tundra–taiga boundary in Eurasia. *Geographical Review*, **60**, 548–54.

Kroeber, A.L. (1916). *Floral relations among the Galapagos Islands*. University of California, Berkeley, Publications in Botany, **16**, 199–220.

Kruskal, J.B. and M. Wish (1978). *Multidimensional scaling*. Sage Publications, Beverly Hills.

Kutzbach, J.E. and F.A. Street-Perrot (1985). Milankovitch forcing of fluctuations in the level of tropical lakes from 10 to 0 kyr BP. *Nature*, **317**, 130–4.

Lack, D. (1976). *Island biology*. Blackwell, Oxford.

Lamb, H.H. (1966). Climate in the 1960s. Changes in the world's wind circulation reflected in prevailing temperatures, rainfall patterns and the levels of the African lakes. *Geographical Journal*, **132**, 183–212.

— (1979). Discussion in Rainey *et al. Philosophical Transactions of the Royal Society of London*, B **287**, 340–2.

Lance, G.N. and W.T. Williams (1966). A generalized strategy for computer classifications. *Nature*, **212**, 218.

— (1967). A general theory of classificatory sorting strategies. I. Hierarchical systems. *Computer Journal*, **9**, 373–80.

Lande, R. (1976). Natural selection and random drift in phenotypic evolution. *Evolution*, **30**, 314–34.

Levin, D.A. (1976). Alkaloid-bearing plants: an ecogeographic perspective. *American Naturalist*, **110**, 261–84.

— and S.W. Funderburg (1979). Genome size in angiosperms: temperate versus tropical species. *American Naturalist*, **114**, 784–95.

Lincoln, G.A. (1975). Bird counts either side of Wallace's Line. *Journal of Zoology of London*, **177**, 349–61.

Lindroth, C.H. (1949). *Die fennoskandischen Carabiden*. Vol. 3. Allgemeiner Teil. Goteborgs Kungl. Vetenskaps- och Vitterhets-sammhalles Handlingar, B. Stockholm.

— (1957). *The faunal connections between Europe and North America*. Almquist and Wicksall, Stockholm.

Lough, J.M. *(1980)*. *West African rainfall variations and tropical Atlantic sea surface temperatures. Climatic Monitoring*, 9, 150–7.

—, T.M.L. Wigley and J.P. Palutikof (1983). Climate and climate impact scenarios for Europe in a warmer world. *Journal of Climatology and Applied Meteorology*, 22, 1673–84.

Lutz, F.E. (1921). Geographic average, a suggested method for the study of distribution. *American Museum Novitates*, 5, 1–7.

— (1922). Altitude in Colorado and geographical distribution. *Bulletin of the American Museum of Natural History*, 10, 335–66.

Lynch, J.F. and N.K. Johnson (1974). Turnover and equilibrium in insular avifaunas, with special reference to the California Channel Islands. *Condor*, 76, 370–84.

MacArthur, R.H. (1972). *Geographical ecology*. Harper and Row, New York.

— and E.O. Wilson (1967). *The theory of island biogeography*. Princeton University Press, Princeton.

Marshall, J.K. (1968). Factors limiting the survival of *Corynephorus canescens* (L.) Beauv. in Great Britain at the northern edge of its distribution. *Oikos*, 19, 206–16.

Martin, P.S. (1967). Prehistoric overkill. In *Pleistocene extinctions: the search for a cause*, eds. P.S. Martin and H.E. Wright, pp. 75–120. Yale University Press, New Haven.

May, R.M. (1975). Patterns of species abundance and diversity. In *Ecology and evolution of communities*, eds. M.L. Cody and J.M. Diamond, pp. 81–120. Harvard University Press, New York.

Mayr, E. (1956). Geographical character gradients and climatic adaptation. *Evolution*, 10, 105–8.

— (1957). Species concepts and definitions. In *The species problem*, ed. E. Mayr, pp. 371–88. American Association for the Advancement of Science, Publication 50, Washington D.C.

— (1963). *Animal species and evolution*. Harvard University Press, Cambridge, Mass.

— (1965a). What is a fauna? *Zoologisches Jahrbuch der Systematik*, 92, 473–86.

— (1965b). The nature of colonizations in birds. In *The genetics of colonizing species*, eds. H.G. Baker and G.L. Stebbins, pp. 29–47. Academic Press, New York.

— (1982). *The growth of biological thought*. Harvard University Press, Cambridge, Mass.

McCormick, J.F. and R.B. Platt (1980). Recovery of an Apallachian forest following the Chestnut Blight or Catherine Keever – you were right! *American Midland Naturalist*, 104, 264–73.

McCoy, E.D., S.S. Bell and K. Walters (1986). Identifying boundaries along environmental gradients. *Ecology*, 67, 749–59.

McCoy, E.D. and K.L. Heck (1987). Some observations on the use of taxonomic similarity in large-scale biogeography. *Journal of Biogeography*, 14, 79–87.

McKillup, S.C., P.G. Allen and M.A. Skewes (1988). The natural decline of an introduced species following its initial increase in abundance; an explanation for *Ommatoiulus moreletii* in Australia. *Oecologia*, 77, 339–42.

McNab, B.K. (1971). On the ecological significance of Bergmann's Rule. *Ecology*, 52, 845–54.

Means, D.B. and D. Simberloff (1987). The peninsula effect: habitat-correlated species decline in Florida's herpetofauna. *Journal of Biogeography*, 14, 551–68.

Menozzi, P., A. Piazza and L.L. Cavalli-Sforza (1978). Synthetic maps of human gene frequencies in Europeans. *Science*, **201**, 786–92.

Messenger, P.S. (1970). Bioclimatic inputs to biological control and pest management programs. In *Concepts of pest management*, eds. R.L. Rabb and F.E. Guthrie, Raleigh.

Michelbacher, A.E. and J. Leighly (1940). The apparent climatic limitations of the Alfalfa weevil in California. *Hilgardia*, **13**, 101–39.

Mitchell, J.M. (1976). An overview of climatic variability and its causal mechanisms. *Quaternary Research*, **6**, 481–93.

— (1979). Evidence of a 22-year rhythm of drought in the western United States related to the Hale solar cycle since the 17th century. In *Solar-terrestrial influences on weather and climate*, eds. B.M. Cormack and T.A. Selinga, pp. 125–43. Reidel, Dordrecht.

Mooney, H.A. and E.L. Dunn (1970). Photosynthetic systems of Mediterranean climate shrubs and trees of California and Chile. *American Naturalist*, **104**, 447–53.

Mueller-Dombois, D., K.W. Bridges and H.L. Carson (1981). *Island ecosystems*. Hutchinson and Ross, Stroudsburg.

Neilson, R.P. (1986). High-resolution climatic analysis and southwest biogeography. *Science*, **232**, 27–34.

— and L.H. Wullstein (1983). Biogeography of two southwest American oaks in relation to atmospheric dynamics. *Journal of Biogeography*, **10**, 275–97.

Nix, H.A. and J.D. Kalma (1972). Climate as a dominant control in the biogeography of Northern Australia and New Guinea. In *Bridge and barrier—The natural and cultural history of Torres Strait*, ed. D. Walker, pp. 61–91. Australian Natural University Department of Biogeography, Geomorphology Publication BG/3.

Packard, G.C. (1967). House sparrows: evolution of populations from the Great Plains and Colorado Rockies. *Systematic Zoology*, **16**, 73–89.

Parry, M.L. (1976). County maps as historical sources. *Scottish Studies*, **19**, 15–26.

— (1978). *Climatic change, agriculture and settlement*. Dawson and Archon, Folkestone.

— and T.R. Carter (1985). The effect of climatic variations on agricultural risk. *Climatic Change*, **7**, 95–110.

—, T.R. Carter and N.T. Konijn (1988). *The impact of climatic variations on agriculture. Vol. 1. Assessments in cool temperate and cold regions*. Kluwer, Dordrecht.

Patterson, B.D. (1984). Mammalian extinction and biogeography in the southern Rocky Mountains. In *Extinctions*, ed. M.H. Nitecki, pp. 247–93. University of Chicago Press, Chicago.

Pearcy, R.W. and J. Ehleringer (1984). Comparative ecophysiology of C3 and C4 plants. *Plant, Cell and Environment*, **7**, 1–13.

Peters, J.A. (1955). The use and misuse of the biotic province concept. *American Naturalist*, **89**, 21–8.

Peters, R.H. (1983). *The ecological implications of body size*. Cambridge University press, Cambridge.

Phillips, D.L. (1978). Polynomial ordination: a field and computer simulation testing of a new method. *Vegetatio*, **37**, 129–40.

Phipps, J.B. (1975). Bestblock: optimizing grid size in biogeographic studies. *Canadian Journal of Botany*, **53**, 1447–52.

Pianka, E.R. (1966). Latitudinal gradients in species diversity: a review of concepts. *American Naturalist*, **100**, 33–46.

Pielou, E.C. (1975). *Ecological diversity*. Wiley, New York.

— (1977). The latitudinal spans of seaweed species and their patterns of overlap. *Journal of Biogeography*, **4**, 299–311.

— (1979a). Interpretation of paleoecological similarity matrices. *Paleobiology*, 5, 435–443.

— (1979b). *Biogeography*. Wiley, New York.

— (1981). The usefulness of ecological models: a stock-taking. *Quarterly Review of Biology*, 56, 17–31.

— (1984). *The interpretation of ecological data*. Wiley, New York.

Pigott, C.D. (1968). Biological flora of the British Isles. *Cirsium acaulon* (L.) Scop. *Journal of Ecology*, 56, 597–612.

— (1974). The response of plants to climate and climatic change. In *The flora of a changing Britain*, ed. F. Perring, pp. 32–44. Classey, Hampton.

— (1975). Experimental studies on the influence of climate on the geographical distribution of plants. *Weather*, 30, 82–90.

— (1981). Nature of seed sterility and natural regeneration of *Tilia cordata* near its northern limit in Finland. *Annales Botanici Fennici*, 18, 255–63.

— and J.P. Huntley (1981). Factors controlling the distribution of *Tilia cordata* at the northern limits of its geographical range. III. Nature and causes of seed sterility. *New Phytology*, 87, 817–39.

— and S.M. Walters (1954). On the interpretation of the discontinuous distributions shown by certain British species of open habitats. *Journal of Ecology*, 42, 95–116.

Pratt, D.M. (1943). Analysis of population development in *Daphnia* at different temperatures. *Biological Bulletin*, 85, 116–40.

Prentice, H.C. (1986). Climate and clinal variation in seed morphology of the white campion, *Silene latifolia* (Caryophyllaceae). *Biological Journal of the Linnean Society*, 27, 179–89.

Preston, F.W. (1948). The commonness, and rarity, of species. *Ecology*, 29, 254–83.

— (1962). The canonical distribution of commonness and rarity. *Ecology*, 43, 185–215, 410–32.

Prince, S.D. and R.N. Carter (1985). The geographical distribution of prickly lettuce (*Lactuca serriola*). III. Its performance in transplant sites beyond its distribution limit in Britain. *Journal of Ecology*, 73, 49–64.

Prins, H.H.T. (1988). Plant phenology patterns in Lake Manyara National Park, Tanzania. *Journal of Biogeography*, 15, 465–80.

Pschorn-Walcher, H. (1954). Die 'Zunahme' der Schadlingsauftreten im Lichte der rezenten Klimagestaltung. *Anzeiger für Schädlingskunde*, 27, 89–91.

Raffauf, R.F. (1970). *A handbook of alkaloids and alkaloid-bearing plants*. Wiley, New York.

Rainey, R.C. (1963). Meteorology and the migration of desert locusts. *Anti-Locust Memoirs*, 7, 1–115.

— (1978). The evolution and ecology of flight: the 'oceanographic' approach. In *Evolution of insect migration and diapause*, ed. H. Dingle, pp. 33–48. Springer, Heidelberg.

—, E. Betts and A. Lumley (1979). The decline of the desert locust plague in the 1960s: control operations or natural causes? *Philosophical Transactions of the Royal Society of London*, B 287, 315–44.

Ralls, K. and P.H. Harvey (1985). Geographic variation in size and sexual dimorphism of North American weasles. *Biological Journal of the Linnean Society*, 25, 119–67.

Rapoport, E.H. (1982). *Areography*. Pergamon Press, Oxford.

Raunkiaer, C. (1934). *The life forms of plants and statistical plant geography*. Clarendon Press, Oxford.

Raup, D.M. and R.E. Crick (1979). Measurement of faunal similarity in paleontology. *Journal of Paleontology*, 53, 1213–27.

— and D. Jablonski, eds. (1986). *Patterns and processes in the history of life*. Springer, Berlin.

Reader, R.J. (1983). Using heat sum models to account for geographic variation in the floral phenology of two ericacious shrubs. *Journal of Biogeography*, **10**, 47–64.

Reddingius, J. and P. J. Den Boer (1970). Simulation experiments illustrating stabilization of animal numbers by spreading of risk. *Oecologia*, **5**, 240–84.

Regal, P.J. (1982). Pollination by wind and animals: ecology of geographic patterns. *Annual Review of Ecology and Systematics*, **13**, 497–524.

Reynolds, C.S. (1984). *The ecology of freshwater phytoplankton*. Cambridge University Press, Cambridge.

Reynolds, J.C. (1985). Details of the geographic replacement of the red squirrel (*Sciurus vulgaris*) by the grey squirrel (*Sciurus carolinensis*) in eastern England. *Journal of Animal Ecology*, **54**, 149–62.

Rice, J. and R.J. Belland (1982). A simulation study of moss floras using Jaccard's coefficient of similarity. *Journal of Biogeography*, **9**, 411–9.

Robertson, G.W. (1973). Development of simplified agroclimatic procedures for assessing temperature effects on crop development. In *Plant response to climatic factors*, ed. R.O. Slatyer. UNESCO, Paris.

Rogers, D.J. (1979). Tsetse population dynamics and distribution: a new analytical approach. *Journal of Animal Ecology*, **48**, 825–49.

— and S.E. Randolph (1986). Distribution and abundance of tsetse flies (*Glossina* spp.). *Journal of Animal Ecology*, **55**, 1007–25.

Rothmahler, W. (1955). *Allgemeine Taxonomie und Chorologie der Pflanzen*. Fischer, Jena.

Rowell, A.J., D.J. McBride and A.R. Palmer (1973). Quantitative study of Trempealeauian (Latest Cambrian) trilobite distribution in North America. *Geological Society of America Bulletin*, **84**, 3429–42.

Ruddiman, W.F. and A. McIntyre (1981). The North Atlantic Ocean during the last deglaciation. *Palaeogeography, Palaeoclimatology, Palaeoecology*, **35**, 145–214.

Ryrholm, N. (1988). An extralimital population in a warm climatic outpost: the case of the moth *Idaea dilutaria* in Scandinavia. *International Journal of Biometeorology*, **32**, 205–16.

Salisbury, E.J. (1926). The geographical distribution of plants in relation to climatic factors. *Geographical Journal*, **57**, 312–35.

— (1942). *The reproductive capacity of plants*. Bell, London.

Schenck, H.G. and A.M. Keen (1936). Marine molluscan provinces of western North America. *Proceedings of the American Philosophical Society*, **76**, 921–38.

Schmidt-Nielsen, K. (1984). *Scaling*. Cambridge Univerity Press, Cambridge.

Schoener, T.W. (1987). The geographical distribution of rarity. *Oecologia*, **74**, 161–73.

Schuster, R.M. (1976). Plate tectonics and its bearing on the geographical origin and dispersal of the angiosperms. In *The origin and early evolution of angiosperms*, ed. C.B. Beck. Columbia University Press, New York.

Schwerdtfeger, F. (1968). *Oekologie der Tiere. Vol. 2. Demoekologie*. Parey, Hamburg.

Sclater, P.L. (1858). On the general geographical distribution of the members of the class Aves. *Journal of the Proceedings of the Linnean Society (London) (Zoology)*, **2**, 130–45.

Shannon, C.E. and W. Weaver (1949). *The mathematical theory of communication*. Urbana.

Sheldeshova, G.G. (1967). Ecological factors determining distribution of the codling moth *Laspeyresia pomonella* L. (Lep., Tortricidae) in the Northern and Southern Hemispheres. *Entomological Review*, **46**, 349–61.

Short, J., G. Caughley, D. Grice and B. Brown (1983). The distribution and abundance of Kangaroos in Western Australia. *Australian Wildlife Research*, **10**, 435–51.

Simberloff, D.S. (1980). A succession of paradigms in ecology: essentialism to materialism and probabilism. *Synthese*, **43**, 3–39.

— (1981). Community effects of introduced species. In *Biotic crises in ecological and evolutionary time*, ed. M.H. Nitecki, pp. 53–81. Academic Press, New York.

— (1987). Introduced species: a biogeographic and systematic perspective. In *Ecology of biological invasions of North America and Hawaii*, eds. H.A. Mooney and J.A. Drake, pp. 3–26. Springer, New York.

Simoons, F.J. (1978). The geographic hypothesis of lactose malabsorption. *American Journal of Digestive Disease*, **23**, 963–80.

— (1981). Celiac disease as a geographic problem. In *Food, nutrition and evolution*, eds. D.N. Walcher and N. Kretchmer. Masson Publishing, New York.

Simpson, G.G. (1960). Notes on the measurement of faunal resemblance. *American Journal of Science*, **258A**, 300–11.

— (1961). *Principles of animal taxonomy*. Columbia University Press, New York.

— (1964). Species density of North American recent mammals. *Systematic Zoology*, **13**, 57 –73.

— (1977). Too many lines: the limits of the oriental and Australian zoogeographic regions. *Proceedings of the American Philosophical Society*, **121**, 107–20.

— (1980). *Why and how*. Pergamon Press, Oxford.

Sjögren, P., J. Elmberg and S.-A. Berglind (1988). Thermal preference in the pool frog *Rana lessonae*: impact on the reproductive behaviour of a northern fringe population. *Holarctic Ecology*, **11**, 178–84.

Skellam, J.C. (1951). Random dispersal in theoretical populations. *Biometrika*, **38**, 196–218.

Skre, O. (1979). The regional distribution of vascular plants in Scandinavia with requirements for high summer temperatures. *Norwegian Journal of Botany*, **26**, 295–318.

Slud, P. (1976). Geographic and climatic relationships of avifaunas with special reference to comparative distribution in the neotropics. *Smithsonian Contributions to Zoology*, **212**, 1–149.

Sneath, P.H.A. and R.R. Sokal (1973). *Numerical taxonomy*. Freeman, San Francisco.

Snijders, T.A.B. (1984). Enumeration and simulation methods for 0–1 matrices with given marginals. *Heymans Bulletins Psychologische Instituten, Rijksuniversiteit Groningen (The Netherlands)*, Report HB-89-949-EX.

Sørensen, T. (1941). *Temperature relations and phenology of the northeast Greenland flowering plants*. Reitzels Forlag, Kopenhagen.

Stanley, S.M. (1986). *Extinction*. Freeman, New York.

Stebbins, G.L. (1971). *Chromosomal evolution in higher plants*. Arnold, London.

Stehli, F.G. (1965). Paleontologic technique for defining ancient ocean currents. *Science*, **145**, 943–46.

— (1968). Taxonomic diversity gradients in pole location: the recent model. In *Evolution and environment*, ed. E.T. Drake, pp. 163–227. Yale University Press, New Haven.

— and C.E. Helsley (1963). Paleontologic technique for defining ancient pole positions. *Science*, **142**, 1057–9.

— and J.W. Wells (1971). Diversity and age patterns in hermatypic corals. *Systematic Zoology*, **20**, 115–26.

Stern, R.D. (1980). The calculation of probability distributions for models of daily precipitation. *Archives fur Meteorologie, Geophysik, und Bioklimatologie, Ser. B*, **28**, 137–47.

Stowe, L.G. and J.L. Brown (1981). A geographic perspective on the ecology of compound leaves. *Evolution*, **35**, 818–21.

Strauss, R.E. (1982). Statistical significance of species clusters in association analysis. *Ecology*, **63**, 634–9.

Strong, D.R. (1979). Biogeographic dynamics of insect-host plant communities. *Annual Review of Entomology*, **24**, 89–119.

—, J.H. Lawton and R. Southwood (1984). *Insects on plants*. Blackwell, Oxford.

Taylor, J.D. and C.N. Taylor (1977). Latitudinal distribution of predatory gastropods in the eastern Atlantic shelf. *Journal of Biogeography*, **4**, 73–81.

Taylor, L.R. and R.A.J. Taylor (1983). Insect migration as a paradigm for animal movement. In *The ecology of animal movement*, eds. I.R. Swingland and P.J. Greenwood, pp. 184–214. Oxford University Press, Oxford.

Teeri, J.A. (1979). The climatology of the C4 photosynthetic pathway. In *Topics in plant population biology*, eds. O.T. Solbrig, S. Jain, G.B. Johnson and P.H. Raven, pp. 356–74. Columbia University Press, New York.

—, L.G. Stowe and D.A. Murawski (1978). The climatology of succulent plant families: Cactaceae and Crassulaceae. *Canadian Journal of Botany*, **56**, 1750–8.

Ter Braak, C.J.F. (1986). Interpreting a hierarchical classification with simple discriminant functions: an ecological example. In *Data analysis and informatics*, ed. E. Diday. Amsterdam.

— and I.C. Prentice (1988). Theory of gradient analysis. *Advances in Ecological Research*, **18**, 271–317.

Thompson, P.A. (1970). Germination of species of Caryophyllaceae in relation to their geographical distribution in Europe. *Annals of Botany*, **34**, 427–49.

— (1973). Seed germination in relation to ecological and geographical distribution. In *Taxonomy and ecology*, ed. K.H. Heywood, pp. 93–119. Academic Press, London.

— (1975). Characterization of the germination response of *Silene dioica* (L.) Clairv. populations in Europe. *Annals of Botany*, **39**, 1–19.

Thompson, V. (1988). Parallel colour form distributions in European and North American populations of the spittlebug *Philaenus spumarius* (L.). *Journal of Biogeography*, **15**, 507–12.

Thornton, I.W.B. (1983). Vicariance and dispersal: confrontation or compatibility? *GeoJournal*, **7**, 557–64.

Turner, J.R.G., J.J. Lennon and J.A. Lawrenson (1988). British bird species distributions and the energy theory. *Nature*, **335**, 539–41.

Uchijima, Z. (1976). Long-term change and variability of air temperature above 10°C in relation to crop production. In *Climate change and food production*, eds. K. Takahashi and M.M. Yoshiro, pp. 217–29. University of Tokyo Press, Tokyo.

Underwood, A.T. (1978). The detection of non-random patterns of distribution of species along a gradient. *Oecologia*, **36**, 317–26.

Valentine, J.W. (1968). The evolution of ecological units above the population level. *Journal of Paleontology*, **42**, 253–67.

Van Balgooy, M.M.J. (1971). Plant-geography of the Pacific as based on a census of phanerogam genera. *Blumea, Suppl.* **6**, 1–222.

Van Beurden, E. (1981). Bioclimatic limits to the spread of *Bufo marinus* in Australia: a baseline. *Proceedings of the Ecological Society of Australia*, **11**, 143–9.

Van den Bosch, F., R. Hengeveld, J.A.J. Metz and A.J. Verkaijk (1990). A new method for analysing animal range expansion. (In preparation.)

Van den Hoek, C. (1975). Phytogeographic provinces along the coasts of the northern Atlantic Ocean. *Phycologia*, **14**, 317–30.

— (1982*a*). Phytogeographic distribution groups of benthic marine algae in the North Atlantic Ocean: a review of experimental evidence from life history studies. *Helgolander wissenschaftichen Meeresuntersuchungen*, **35**, 153–214.

— (1982*b*). Distribution groups of benthic marine algae in relation to the temperature regulation of their life histories. *Biological Journal of the Linnean Society*, **18**, 81–144.

Van Valen, L. (1965). Morphological variation and width of the ecological niche. *American Naturalist*, **94**, 377–90.

Van Zeist, W. (1969). Reflections on prehistoric environments in the Near East. In *The domestication and exploitation of plants and animals*, eds. J. Ucko and G.W. Dimbleby, pp. 35–46. Duckworth, London.

Vermeij, G.J. (1978). *Biogeography and adaptation*. Harvard University Press, Cambridge, Mass.

— (1982). Phenotypic evolution in a poorly dispersing snail after arrival of a predator. *Nature*, **299**, 349–50.

Vinogradova, N.G. (1959). The zoogeographical distribution of the deep-water bottom fauna in the abyssal zone of the ocean. *Deep-Sea Research*, **5**, 205–8.

Walker, D. and J.R. Flenley (1979). Late Quaternary vegetational history of the Enga Province of Upland Papua New Guinea. *Philosophical Transactions of the Royal Society of London*, B **286**, 265–344.

Waloff, Z. (1966). The upsurges and recessions of the Desert Locust plague: an historical survey. *Anti-Locust Memoir*, **8**, 1–111.

Walter, H. and H. Straka (1970). *Einführung in die Phytologie, III*, 2, *Arealkunde*, 2nd. ed. Ulmer Verlag, Stuttgart.

— and E. Walter (1953). Einige allgemeine Ergebnisse unserer Forschungsreise nach Südwestafrika 1952/53: Das Gesetz der relativen Standortskonstanz; das Wesen der Pflanzengesellschaften. *Berichte des deutschen botanischen Gesellschafts*, **66**, 228–36.

Wang, J.Y. (1960). A critique of the heat unit approach to plant response studies. *Ecology*, **41**, 785–90.

— (1963). *Agricultural meteorology*. Milwaukee.

Warnecke, G. (1936). Ueber die Konstanz der oekologischen Valenz einer Tierart als Voraussetzung fur zoogeographische Untersuchungen. *Entomologische Rundschau*, **53**, 203–32.

Webb, L.J. (1959). Physiognomic classification of Australian rain-forests. *Journal of Ecology*, **47**, 551–70.

Webb, T. and T.M.L. Wigley (1985). What past climates can indicate about a warmer world. In *Projecting the climatic effects of increasing carbon dioxide*. Report DOE/ER-0237 of the U.S. Dept. of Energy, pp. 239–57.

Webster, R. (1977). *Quantitative and numerical methods in soil classification and survey*. Oxford University Press, Oxford.

Wellington, W.G. (1952). Air-mass climatology of Ontario north of Lake Huron and Lake Superior before outbreaks of the spruce budworm, *Choristoneura fumiferana* (Clem.), and the forest tent caterpillar, *Malacosoma disstria* Hbn. (Lepidoptera: Tortricidae; Lasiocampidae). *Canadian Journal of Zoology*, **30**, 114–27.

— (1954). Atmospheric circulation processes and insect ecology. *Canadian Entomologist*, **86**, 312–33.

—, J.J. Fettes, K.B. Turner and R.M. Belyea (1950). Physical and biological indicators of the development of outbreaks of the spruce budworm *Choristoneura fumiferana* (Clem.) (Lepidoptera: Tortricidae). *Canadian Journal of Research*, D **28**, 308–31.

Wentworth, T.R. (1983). Distribution of C4 plants along environmental and compositional gradients in southeastern Arizona. *Vegetatio*, **52**, 21–34.

White, M.J.D. (1978). *Modes of speciation*. Freeman, San Francisco.

Whitmore, T.C., ed. (1981). *Wallace's Line and plate tectonics*. Oxford University Press, Oxford.

Whittaker, R.H. (1967). Gradient analysis of vegetation. *Biological Review*, **42**, 207–64.

Wiens, J.A. (1984). On understanding a non-equilibrium world: myth and reality in community patterns and processes. In *Ecological communities: conceptual issues and the evidence*, eds. D.R. Strong, D. Simberloff, L.G. Abele and A.B. Thistle, pp. 439–57. Princeton University Press, Princeton.

Wigley, T.M.L. (1985). Impact of extreme events. *Nature*, **316**, 106–7.

Williams, C.B. (1947). The logarithmic series and the comparison of island floras. *Proceedings of the Linnean Society (London)*, **158**, 104–8.

—— (1949). Jaccard's Generic Coefficient and Coefficient of Floral Community, in relation to the logarithmic series and the Index of Diversity. *Annals of Botany*, **13**, 53–8.

Williams, P.H. (1988). Habitat use by bumblebees (*Bombus* spp.). *Ecological Entomology*, **13**, 223–37.

Williamson, M.H. (1981). *Island populations*. Oxford University Press, Oxford.

Willis, J.C. (1922). *Age and area*. Cambridge University Press, Cambridge.

Wilsie, C.P. (1962). *Crop adaptation and distribution*. Freeman, San Francisco.

Wilson, E.O. and W.L. Brown (1953). The subspecies concept and its taxonomic application. *Systematic Zoology*, **2**, 97–111.

Wishart, D. (1969). An algorithm for hierarchical classifications. *Biometrics*, **25**, 165–70.

Wolda, H. (1963). Natural populations of the polymorphic landsnail *Cepaea nemoralis* (L.). *Archives Neerlandaises de Zoologie*, **15**, 381–471.

—— (1981). Similarity indices, sample size and diversity. *Oecologia*, **50**, 296–302.

Wolfe, J.A. (1978). A paleobotanical interpretation of Tertiary climates in the Northern Hemisphere. *American Scientist*, **66**, 694–703.

Woodson, R.E. (1964). The geography of flower color in butterfly weed. *Evolution*, **18**, 143–63.

Woodward, S.P. (1856). *Manual of the Mollusca: a treatise on recent and fossil shells*. London.

Wright, D.H. (1983). Species-energy theory: an extension of species-area theory. *Oikos*, **41**, 496–506.

Wright, H.R. (1977). Quaternary vegetation history – some comparisons between Europe and America. *Annual Review of Earth and Planetary Sciences*, **5**, 123–58.

Wulff, E.V. (1943). *An introduction to historical plant geography*. Chronica Botanica Company, Waltham, Mass.

Wyndham, E. (1986). Length of bird's breeding seasons. *American Naturalist*, **128**, 155–64.

Yarish, C., A.M. Breeman and C. Van den Hoek (1986). Survival strategies and temperature responses in seaweeds belonging to different biogeographic distribution groups. *Botanica Marina*, **29**, 215–30.

Yoshino, M.M., T. Horie, H. Seino, H. Tsujii, T. Uchijima and Z. Uchijima (1988). The effects of climatic variations on agriculture in Japan. In *The impact of climatic variations on agriculture. Vol. 1. Assessments in cool temperate and cold regions*, eds. M.L. Parry, T.R. Carter and N.T. Konijn, pp. 725–868. Kluwer, Dordrecht.

Zohari, D. (1965). Colonizer species in the wheat group. In *The genetics of colonizing species*, eds. H.G. Baker and G.L. Stebbins, pp. 404–23. Academic Press, New York.

— (1969). The progenitors of wheat and barley in relation to domestication and agricultural dispersal in the Old World. In *The domestication of plants and animals*, eds. P.J. Ucko and G.W. Dimbleby, pp. 47–66. Duckworth, London.

— (1970). Centres of diversity and centres of origin. In *Genetic resources in plants – their exploration and conservation*, eds. O.H. Frankel and E. Bennett, pp. 33–42. Blackwell, Oxford.

Author index

Species index

Subject index

abiotic factors 162, 163
adaptability, genetic 121
advancing-wave model 123
air current 211
air mass origin 200
algorithm
  agglomerative 44
  divisive 44, 48
  generalized 45
alkaloid 114
approach
  biological 8
  deductive 13
  evolutionary 10
  individualistic 11
  inductive 13
  multidimensional 188
  non-dimensional 188
  oceanic 3
  qualitative 25
  quantitative 27
avian diversity 87

Baker's Law 135
balance of nature 186
biogeographical laws 9
biogeography 20
  dynamic 1, 7
  island 15
  vicariance 15, 31, 194
biological theories 8
biome 201
biotic factors 162
blocking, atmospheric 212
bridgehead 192
Brownian movement 170

catastrophe theory 193
catastrophes 194
centre

of diversity 94
of origin 94
chaining 47
chamaephyte 105
circumpolar vortex 174, 211
classification
  criteria 32
  hierarchical 59
  model 32
  monothetic 28, 32, 62
  non-hierarchical 60
  polythetic 28, 32, 59, 62
climatic indicators 166
climatic reconstruction 166
clustering
  agglomerative 44
  B–k 61
  centroid 47
  complete-linkage 47
  divisive 44, 48
  furthest-neighbour 47
  intensity coefficient 46
  k- 61
  median 47
  nearest-neighbour 47
  single-linkage 46
  sum-of-squares 47
  u-diametric 61
  Ward's method 47, 48
  weighted centroid 47
coefficient
  of concordance 49
  of cophenetic correlation 49
  diversity 34
compensation
  altitudinal 134
  geographical 134
  habitat 134
  seasonal 134
competition 135, 165
compound leaves 104

246

Printed in the United States
By Bookmasters